MORE MOGOLLON MONSTER, ARIZONA'S BIGFOOT

SUSAN FARNSWORTH AND MITCH WAITE

DEDICATION

To all those serious and hard-working Bigfoot Researchers of the past, present, and future. Don't let anyone get you down, or discourage you from pursuing the truth. Basically, nothing has been proven, only theory. This makes you as much an expert in the field as any of the big names of Bigfoot Research. Keep on Bigfooting!

CONTENTS

TABLE OF CONTENTS (CONTINUED)

ACKNOWLEDGMENTS

WE WOULD LIKE TO ACKNOWLEDGE ALL
OUR GOOD FRIENDS IN THE BIGFOOT WORLD
THAT HAVE INSPRIED US TO CONTINUE OUR
BIGFOOT RESEARCH.

INTRODUCTION:

Mitch and I have decided to combine our efforts into one book The first half of the book will deal with what I do best, and that is the collection of Mogollon Monster tales and stories from around the campfire.

These stories as told by the Local People of Northern Arizona, and can't be verified. After all, they are campfire stories. However, many of the stories are based on the truth, and should be recorded for posterity and used for comparison to any scientific findings that Bigfoot researchers should make.

The second half of the book consists of journal records of Mitch Waite and his exploits as he searched the Mogollon Rim for Bigfoot. Mitch and I are fairly new to the Bigfoot field research field. But, we soon found out this field of work is very addictive. We were hooked and are fully engaged in research of five "Hot Spot" areas within the Mogollon Rim of Arizona.

Back in 1966, I published my first book on the Mogollon Monster, Arizona's Bigfoot with the subtitle Stories Of The Mogollon Monster As Told By The Local People of Northern Arizona. It was done on a small scale because I was an unpublished author, and the market was unsure. The publisher, Southwest Publications, agreed to publish a limited number of books to test the market. They were sold through major book store chains via the ISBN system. They eventually all sold, but the demand didn't justify a larger reprint.

I kept a couple of cases of books passed them out to friends and relatives. Ten years later I decided to try remarketing the remaining books. Suddenly, they took off. I was amazed when I GOOGLE'd the Mogollon Monster and found my book or my name on almost every page.

I started getting email on the subject. A lot of people hailed me as being the expert on the Mogollon Monster, while others started calling me a charlatan and fraud claiming that I was trying to cash in on folk lore of Washington, Oregon and California—Those places with "real" Bigfoot creatures.

I would like to state I am neither an expert nor fraud. I did not have proof of the existence or nonexistence of any such creature, especially the Mogollon Monster. However, I was fascinated by the stories told by the local people who live on the Mogollon Rim.

There was enough anecdotal evidence to indicate there was and is something out there, and I started collecting these stories. I believe they are an important part of Arizona's culture and heritage. It was after sharing many of these stories, that I was urged by many to compile them into a book. That is precisely what I did. I would also like to add, I have never been to the Washington, Oregon, or Alaska areas. I have no clue as to their cultural tales of Bigfoot. Therefore, it would be very difficult for me to cash in on the Northwest legends and lore. Perhaps, someone should compile a book of those tales.

I found out there basically two types of opinions. One is you believe, and the other is you don't.

Those who believe usually have some type of experience to relate, while those who don't believe poke fun at those who do. Which makes it very hard to collect stories with more detail than—"I saw it. It was big. It smelled. It scared the crap out of me. It ran away."

Therefore, I decided to take a new approach to the Mogollon Monster legends and lore. Go to the campfires and listen. Don't judge, just compile the stories. And this is what I did. I make no effort to verify the validity of the stories, no cross examination, and no names (if the story teller desires). Under these conditions, people loosen up and talk, and I get wonderfully detailed stories that need to be recorded. If I were to show any skepticism, the story telling would stop, and I would probably soon find myself banned from the local circle.

It is interesting how most professed non-believers, who poke fun at those who believe, spend very little time in the forests on their own. And, they are not familiar with the extremely rough and remote places of Arizona. Something tells me that even the hardest of non-believers get a little nervous when they are in these deep woods alone. These skeptics usually have produced nothing of value, but are very good at providing nonproductive comments, in a mean spirited and contentious manner.

The first part of this book records the stories. Some of the stories come with some supporting data.

There may be a point on the map, a photo, or maybe a sketch. Again, no effort has been made to verify or investigate the authenticity. That is strictly for the experts. I suggest the experts look for commonalities in the tales as a possible indicator as to the true nature of the beast and not waste time trying to nitpick the stories.

It's hard to figure out where to start when you talk Bigfoot verses the Mogollon Monster. I guess the first thing that most people probably know, but don't realize; is we are not talking about one creature, but a group or pod(s). The Bigfoot of Washington, Oregon, and California seem to be darker in color. Their fur is usually described to be a dark brown or even black.

The Mogollon Monster of the Escudilla (eastern Arizona) area seem to be more the color of an Orangutan (burnt orange). Speaking of color, I have heard of white, brown, red, burnt orange, and black Bigfoot.

Could this be an indicator of a difference in species or sub species, or is it like comparing a human blond to a human redhead? I don't know. There also seems to be a change in color as they age. Perhaps this is like the silverback gorillas, as they age, the hair turns silver.

Vocalizations are another area of difference between BF and MM. The western Bigfoot (CA, OR, WA) is more like howler monkeys in their calls. Most sound like moans and wails. Close proximity vocalizations sound much like a drunk Sumo Wrestler cussing out his wife. The verbal tones are strained and low.

The Mogollon Monster, is reported time and time again to sound like a screaming woman who is fighting for her life. Yet, the person reporting the sounds usually declare that the noise was definitely not a mountain lion. And, when in close proximity to humans, there may be low huffs, grunts, and growls. My gut feeling tells me that you may be pushing your luck if you hear the growl. Keep pressing and you may end up with your arms and legs going in directions they shouldn't be pointing.

Recently, we have been able to record some vocalizations, and compare them to other vocalizations around the world. The internet is a powerful tool, and it has allowed us to match several types of vocalizations to be the same across the country. Therefore, there are some differences in vocalizations probably due to the local dialect, and there are some very similar vocalizations which indicates a base language.

Keep in mind, these animals are intelligent, but they are wild creatures trying to survive. They will protect themselves and their young. Give them respect, and don't do anything foolish. They do seem to know a firearm and what it can do. They also display curiosity in non-threatening events. Some may even have a sense of humor. So, if you are lucky enough to have an encounter; don't do anything stupid that may aggravate the situation. Be smart, enjoy, and remember you are one of the lucky few.

Susan Farnsworth

PART 1: MORE MOGOLLON MONSTER TALES
COMPILED BY SUSAN FARNSWORTH

STORY 1

27 October, 1975, Arizona Hunting Area 1, Greens Peak Wilderness

"It's going to be a great day for hunting." Steve commented as he loaded his pack with snacks to last him through the day. "I'm going to work my way down the entrance road a couple of miles and then drop into the wide canyon back up to the peak."

"Feeling ambitious today," Lance commented while he checked his gun. "Remember, if you shoot it, you have to carry it out."

"That's the problem with hunting the Wilderness area." Steve admitted. But, there are a lot of old fire break roads. I can use the deer carrier."

"You really think that thing will work?"

"Sure."

"That still means you have to hike all the way back here to get it, and then all the way back to the animal. Then all the way back here again." Lance laughed.

"Maybe I'll just herd it up here to the truck and then shoot it."

"Good luck with that one." Lance smirked and then added, "I'm going to make my way around the base of the peak and be back here later in the afternoon. I'm too sore to walk much further."

Lance turned and walked away. Within a few moments he had disappeared into the tree line.

This was the second day of the deer hunt. So far all the two hunters had managed to scare up was a small heard of cow elk on the first day. Steve remembered thinking how skittish the elk seemed to be. They acted more like deer. A twig snapped under Steve's boot, and the elk bolted and were gone.

Today was going to be different. Steve knew he was going to see some deer. All he wanted was one good shot and a four point buck. That was all he asked.

The hike was easy going down the old fire break trail. Steve also liked the fact that he could travel fairly fast and very quietly. No twigs to break underfoot, or pine cones to crunch and no rocks to trip over.

He was at his departure point from the road in short order. The sun was starting to make its appearance by cutting through the tall pine trees.

There was a light fog coming up from the canyon. However, the sun was doing its job by burning the fog off before it topped the ridge. But, it gave Steve an eerie feeling as he started down into the wide canyon.

"This is where the deer will be." Steve thought to himself. He took his rifle off safety as he sat down on a rock overlooking the canyon floor below. This was a good spot.

Suddenly, there was a shrill scream from the next ridge. It sounded as if a woman was being attacked. Steve swung his rifle up towards the noise and began to study the ridge through his scope. The scream was followed by the ferocious roar of a bear in battle and followed by more screams.

Steve was completely at a loss as what to do. Was it a bear attack on a woman hunter? Why were there no shots.

It was quite evident, the incident was taking place just over the ridge on the opposite side of the canyon. Quickly as it started, it ended.

Steve spotted the movement first, then he could hear the huffs of air the bear was making as it topped over the ridge and headed down into the canyon in a dead run. The bear was making a hasty retreat away from the battle zone. And still no shots were fired. Was it a fight with another bear? It sure didn't sound like it.

There was no pursuit. Steve waited for the dominate bear to at least top the ridge to verify its victory. But, there was none. Steve could tell by the noise, the fleeing bear had gone down into the wide canyon and turned downstream away from him. At least he was not going to have to confront the fleeing bear. It was leaving the area.

"I should go over the ridge and make sure it wasn't another hunter," Steve thought to himself. "But, what if it is another bear, and it is still hot from the battle?"

Steve reassured himself that his 7mm magnum would be able to stop any bear. He reasoned that he would just stay clear from any ambush points. But, he had to go make sure.

He stood up and started down the hillside. The whole distance was only about five hundred yards, but the closer he got the more leery he became. He stopped short of the top of the ridge to catch his breath and to check his gun. It was loaded and ready to fire.

Steve peered over the ridge carefully. He could see a small hollow on the other side. The ground was messed up. The grass showed marks of a struggle, but there was no body. This made Steve feel much better. He really didn't want to find a human corpse. And, in this area any one hanging on to life probably would not make the trip out to civilization or last long enough for someone to go get help.

Cautiously, he moved over the ridge to inspect the area. Blood was splattered in several places, and long with fur and hair. Some of the hair didn't match the bear. It was long and black. The bear was brown. Wolverine maybe? But, smaller animal wouldn't cause the bear to tear up the ground so much.

Steve noticed an outcropping of rocks about twenty five yards uphill. The seemed to form a natural wall. There was a blood trail in that direction. Again, he had to bolster himself to follow through with checking out the area. It would have been much better to have Lance with him. But, Lance was miles away near the peak.

Steve moved towards the rocks and slowly looked over the rocks into a small pit. It was more of a natural indentation in the earth with the rocks forming one side. The pit area was approximately ten yards in diameter. It was what was in the pit, the surprised him. Six dead doe deer. The carcasses were neatly stacked round the sides of the pit.

Curiosity got the better of Steve. He had to find out what had killed these deer and stacked them in the pit. He moved over the side and into the pit for a closer inspection. The deer seemed fine. No bullet holes or puncture wounds. However, the heads were all missing.

Then it hit him like a ton of bricks. Something had killed these deer by snapping their necks and removing the heads. One deer's blood was still running. The kill was fresh.

Steve was standing in some beast's pantry. This beast was large enough to chase off a bear, and whatever it was, it was not too far away. The bear ran, and the victor would not. That meant it was very close, perhaps tending to its wounds.

He needed to get out of there as quietly as possible. Steve scrambled over the side of the pit and ran down the hillside into the canyon he had just crossed. He jogged across the canyon and headed back to the fire break road. When he finally stopped to catch his breath. His heart was pounding in his chest.

A tree limb snapped under the weight of something behind him. He thought of the elk the day before. He knew why they were skittish, and he knew he had to run. He bolted down the fire break road towards the vehicle. Sounds in the bushes told him he was being followed just out of sight and in the tree line.

Finally he topped the last ridge and could see the vehicles. Lance was near the front of the truck and heard Steve coming down the road. He grabbed his gun and started towards Steve. Steve screamed for Lance to get in the truck and start it up. Lance complied. Steve made the truck and tossed his rifle into the back.

"Hit it!" He commanded. "Get us out of here!"

Lance floored the gas petal and they sped down the road to safety. Their camp stayed abandoned.

MITCH'S COMMENTS:

Again this story has some elements that ring true. I have had the opportunity to visit several deer kills. Some of the things in common with the story is the head is missing. Usually, the front legs are broken or removed at the knees. The heart and lungs are missing, the hide will be skinned back to the hind hocks exposing the entire carcass. Then the meat is stripped from the bones and the skeleton remains intact. Predators of the wild do not do this.

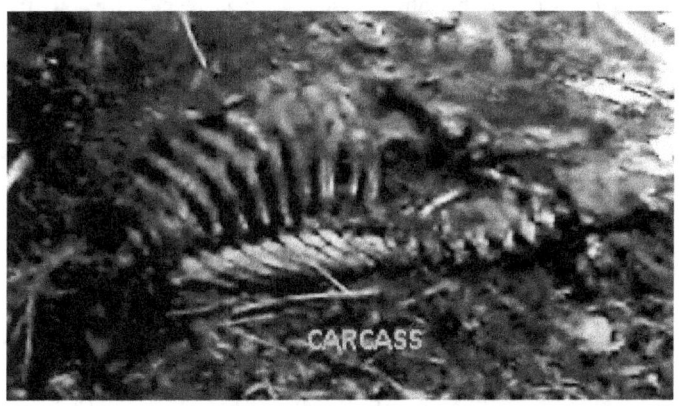

CARCASS

Story 2

LATE SUMMER, 2001 NEAR CHEVELON LAKE.

"I hope this hike is worth it" Alice complained. "What is it, a half mile?"

"Yes, the lake is about a half mile down the trail." Richard tried convince his wife the trek would be good. "While I do some fishing you can pick some blackberries."

"There are blackberries?"

"Yes, I think that is what they are. They are black and look like raspberries."

"You sure they are eatable?"

"Yes, I picked some last Summer and ate them for breakfast."

"Do they have thrones?"

"Yes. So wear long pants and bring your gloves."

The two were making an extra effort to try each other's interests. There was no real trouble between them, but they were trying to make sure it didn't get that way. They had been arguing more than usual and that was when they decided to try to become closer. It was Alice's turn to try one of Richard's hobbies.

It was good to be out and away from the city. They had specifically chosen not to go during Spring break, but wanted to time it where they could have the forest mostly to themselves. Their timing was excellent. There were no other campers at the Chevelon Lake campsite. They pretty much had the entire area to themselves.

They had arrived at the campsite late Monday afternoon, and set up their folding camp trailer before dark. There was no time to look for much fire wood, and the campsite was pretty clean of any burnable wood. They decided not to make a fire, but to cook over their propane stove and call it an early night. Tomorrow would present plenty of time to gather some wood.

Richard was up early. He wanted to be down at the lake's edge to try his new royal nymph fly on the trout. It was a perfect morning for it. But he knew his wife wouldn't budge out of the bed without the tent being warmed and breakfast on the grill.

Alice had different ideas. She wanted to sleep in and enjoy the morning at her own leisurely rate. The bed always seemed to have a human magnet in it during the mornings keeping her attached to the bed. She hated getting up out of a nice warm and comfortable bed into the frigid cold air of the unheated popup tent. It was bad enough there was no bathroom.

Breakfast had come and gone. Richard fixed his famous scrambled eggs and Texas toast. Alice didn't appreciate the cuisine. She managed to eat the scrambled eggs with jalapeño bits and cheese. The toast wasn't really toast. It was slices of bread browned in a frying pan with a scoop of butter to keep it from burning. She wanted more of a continental breakfast with fruit, cereal, and juice. But, they were roughing it.

"It's going to be a beautiful morning", Richard pointed out the morning glow of the sun, which hand not yet made its appearance over the mountains.

It was light enough to walk, and by the time they made the lake it would be perfect for trout fishing. The trail looked like it had been a road, but the rocks made it very rough. A major four wheel drive would be needed to get down to the water's edge. Unfortunately, that was not in their budget. For now, they would have to walk.

Alice stumbled over a few rocks in the trail, and caught herself before falling. She thought about the trail. It was mostly downhill and that was good for now. But later, they would have to climb back up the same rocky trail back to camp. She couldn't figure out why Richard loved the outdoors so much.

Finally, they reached the lake. Richard took a few minutes setting up his tackle while Alice found a nice flat rock to sit on. The sun was just peeking over the ridge and it felt good to sit on the stone and soak it in.

The fish were not biting, and Richard was trying to figure out how to present his flies and lures in such a manner to entice a fish to strike. He wished the lake was not restricted to artificial flies and lures, and he was sure he could reel them in if he could use some salmon eggs. But, he also knew it was his kind of luck if he gave into cheating, there would be a game ranger behind one of the trees ready to write him a ticket. So he continued to cast his flies.

It didn't take long for Alice to get bored. She had no desire to fish. She didn't even like to fish. And, she especially hated the cloud of gnats that had started to form over her rock.

"Where are the berries?" She finally blurted out.

"You have to work your way around the lake to the steep bank over there."

"Ok. I'm going to go get some berries."

"Just stick to the shore pretty close so I can watch you." Richard insisted.

"Why?"

"Because it is never a good idea to wonder off by yourself."

"I can take care of myself."

"Alice." Richard called after his wife. "We are in the high country. Watch for bears."

"Bears?" Alice mumbled. "Bears too?"

"Yes."

"Well, what about you? You don't seem to be worried about bears."

"I'm armed."

"Well, then what do I do if I see a bear?"

"Try to keep as much distance as you can from them. Don't turn and run. Running tells them you are lunch and you know it."

"What about playing dead?"

"That mostly only works if they already have you."

"Climb a tree?"

"Bears can climb a tree better than you," Richard laughed. "But that will works sometimes."

"I heard you can bluff them some times. Make yourself as big as you can and making lots of noise."

"That might work. Especially if it is smaller than you. But you're no Daniel Boone. I wouldn't try grinning down bear."

"How about diving in the lake?"

"That depends on the bear, and if he wants to get wet. They can swim very well. But they don't necessarily like to go under water. The best thing to do is see them first and stay away from them."

"Ok." Alice caved in. "I'll stick to the shore line."

"I'll start working my way over there as I fish. I'll keep you in sight."

Alice started down the shore to the berries. There were no bears. No chipmunks, no deer, there wasn't much of anything but gnats. Finally, she reached the berry bushes. And, there were berries. Loads of them. She immediately started picking them. Half went into her bucket, and the other half into her mouth. They were very good, and the blacker the berry; the sweeter it tasted.

She spotted a bunch of berries on a steep part of the slope and proceeded to work her way over to them. Going was slow because some of the berry bushes had climbed over the rocks and bushes creating a very thick canopy of vegetation in some places.

She made it to the thick patch of berries and began to pick them. While doing so she spotted her husband not more than seventy five yards from her walking in her direction. She also noticed the lake below. If she lost her footing and didn't get tangled up in the thorns she could end up in the lake.

Suddenly, she saw her husband drop his fishing rod, and draw his pistol. He seemed very irritated and was waving his arms at her. He was yelling something.

She heard the huff first, the blast of air go down her collar, and she knew something very big was standing very close to her. So close, its foul breath felt warm on the back of her neck. She heard it growl when she jumped clearing the bushes into the lake. A split second before she hit the water, she heard a shot. She thought she heard a couple more shots when she surfaced. A few moments later she reached the other side of the inlet and was being helped from the water by her husband.

No words were said, but a hasty retreat to the camp ensued. Richard fumbled in his pockets for his keys as they approached the camper. He unlocked the door and shoved Alice inside and gave her the keys.

"Start the truck!" He demanded as he pulled the leveling jacks from the trailer.

Alice complied and slid over to the passenger side. Moments later they were speeding down the dirt road away from the lake. The pop-up tent still in the up position.

"Richard," Alice tried to break his concentration on the road. "Richard. It was only a bear. You can slow down now."

Alice was wondering why Richard was so unnerved. She should have been the one scared stiff.

"Richard. We're safe. We can stop now and put down the top."

Richard seemed to melt just a little. "Alice, that was no bear."

"What was it then."

"That was a monster. The Mogollon Monster!"

"No. There's no such…"

Richard interrupted her. "That was the biggest, hairiest, man-creature I had ever seen. He must have been eating berries in the bush and you walked right in front of him. He reached for you as you jumped. I fired a shot in the air to scare him away, and the whole berry patch came alive."

"Your kidding."

"No. There was at least two big ones and one little one. I just kept firing in the air to scare them away. You didn't see them?"

"No. Just felt the one breathing down my neck." Alice rubbed her neck thinking about how close it must have been.

"I figure if there were that many, there are bound to be more. We're not stopping until we get to the highway."

Alice managed a nervous smile. "That we can agree on."

MITCH'S COMMENTS:

Chevelon Lake has had several sightings through the years. It is very close to the areas of Woods Canyon Lake, Bear Lake, and Knoll Lake. All of these lakes are on the Mogollon Rim and are considered hot spots for Bigfoot activity. Most of the time, the sightings occur on the non-human side of the lakes or near the dams. It might also be noted that these lakes are full of crayfish which are very easy to catch. These crayfish would provide a very good source of protein to any Bigfoot in the area.

The berries are usually ready in the late summer and are plentiful. It makes sense that the Bigfoot would follow the food sources. However, back berries can be found in almost any of the permanent water places such as the lakes, and creeks. These bushes can climb trees and cover rocks and small caves. It would be possible to walk right past something hiding in these places and never see them.

STORY 3

29 October, 2006, Arizona Hunting Unit 22, Matazal Wilderness, in Foothills of Matazal Peak.

It was the third day of deer season. Randy and John were up before dawn each hunting day to prepare for the hunt. Stuffing snacks into their fanny packs and filling their canteens to the brim. So far, it had been a fruitless ritual. They has seen sign, plenty of droppings, but no deer. Today they were going to press further into the wilderness area to go where no man had been for many years. Hopefully, they would see some game.

I've been thinking," Randy commented while he sipped a cup of hot chocolate. "I think we should work down the tributary to Dead Man's Creek. Work our way up the creek to the Matazel Ridge Trail. Then we can come back to camp on the trail."

John was busy fussing with his fanny pack, but stopped long enough to consider the proposal. "You know that is some mighty rough country."

"Yes, but I think we have to or we aren't going to see any deer."

"You're probably right. First shot on opening day sent them deep."

"Yes, they are no dummies. They do seem to know when it's hunting season."

John smiled then commented. "You know we are up against superior beings?"

"Oh?"

"The deer can out smell us, out run us, and see us way before we see them."

"True." Randy agreed.

"They are masters at tactical maneuvers. This is their home."

"True."

"They know every nook and cranny. They can be twenty feet from you and completely vanish into a ravine or over a ridge. The only thing we have on them is a weapon that can reach out several hundred yards when they make a mistake."

"I never thought of it that way." Randy mused at the thought.

"We are definitely outclassed."

The two continued their preparations. The last thing on the list was to load their rifles and head out of camp. Their timing was just about perfect. It was getting light enough to see, and the sun would be at their backs.

They headed down the small tributary working towards the creek. One on each side of the banks. In this manner they could cover both sides of the canyon visually. The game would have to push down the creek, or make an attempt to climb the sides being fully exposed to the hunter's weapons.

Covering nearly a mile, they had not seen anything. No fresh sign. Upon reaching Dead Man's Creek, they turned north traveling up stream. Progress was slow, and in some points they were forced in to the creek bed. On one occasion, John stopped to check for tracks. What he found, struck him as odd. He whistled at Randy to signal him to come and take a look.

"What you got?" Randy spoke in a low whispering tone.

"Someone is in here," John pointed a single footprint. "And they're bare foot."

"What?"

"See for yourself."

"Ok," Randy walked over to take a closer look. "They must have stopped to soak their feet, or take a rock out of their shoe."

"Maybe. Could be."

"They probably sat on that rock there, and took their shoe off." Randy pointed out a rock about the right height for a good stool.

"I suppose you are right." John admitted. "But then again, he wasn't a small person either."

John stuck his foot down next to the footprint. It was quite apparent his foot could fit inside the track, shoe and all with an inch or two to spare.

"That's got to be a size fourteen or better."

"What are you saying," Randy stated sarcastically. "The Mogollon Monster is in this canyon?"

"Never mind." John brushed the comment off. "I take it you don't believe in Bigfoot?"

"No. Never have, and never will."

"Never?"

"It's all a bunch of hooey. Just a story to scare the kids into staying around the camps and not wondering off."

"Are you sure?" John was trying to unravel Randy's nerve.

"Come on." Randy insisted. "You don't think a creature that big could exist today without being discovered? There's never been any bones, never any DNA proof, and no bodies. Even the pictures are non-conclusive. Probably all hoaxes and the camera is always out of focus."

"Well, I do know one thing about the foot print." John stated.

"What?"

"That explains why there is no deer in here. Bigfoot or man, the deer have moved on somewhere else."

The two started walking again. Working their way up the canyon.

It was just before noon the two stopped for lunch. John noticed the tailings of an old mine shaft.

"Hey, there's an old mine."

"So, these mountains are full of them."

"I wonder if there is any gold?" John teased

"Probably not. These old shafts were cut way back in the early nineteen hundreds. Probably nothing left, and that's why it was abandoned."

"Well, let's take a look at the tailings."

The two moved over to the shaft entrance. The tailings were full of rose quartz and had bits of green composite copper. Some of the quartz looked promising.

"I heard the old timers were only interested in the easy gold." John stated.

"Yes, many mines are being reopened using modern technology and are paying out."

"You think this mine might be worthwhile?"

"Maybe if the vein hasn't petered out."

"Well, there is only one way to find out," John opened his fanny pack and pulled out a flashlight.

"I don't think this is a good idea." Randy stated.

"Well you staying out here or coming in?"

"Oh, all right. I'll go with you, but we have to leave the rifles out here."

"Why?"

"Well, for one, you can't use them in there anyway. In these old shafts, almost anything could cause a cave in."

"Ok, but I am taking my pistol."

"Ok."

The two turned on their flashlights and proceeded into the shaft. Inside, they found the remains of an old mining cart track. A few of the rails were still in place. There was no evidence of any unstable rock or previous cave-ins. The shaft went straight back in to the hillside for nearly seventy five feet. Then it took a sharp right hand turn and went another twenty five yards before it narrowed. The shaft only narrowed in height. The width was large enough for both men to crouch.

"What do you think?" John asked Randy as he flashed his light inside the opening. "It looks like it opens up into a room."

John studied the opening for a few moments. Then quietly he spoke. "I think we need to get out of here."

"Why?" Randy asked.

John flashed his light at the floor of the shaft near Randy's feet. Randy was standing next to another huge human like footprint.

"I think our friend might be in there." John whispered as he drew his pistol flipped the safety off.

"What makes you think so?" Randy replied in a hushed tone.

John moved his light to shine at his feet. "Because this puddle I'm standing in is not water."

Randy looked at the wet spot on the cave floor. "Well, then what is it?"

"It's urine, and it's not mine!"

Inside the cave room a rock slid underneath something massive. It was enough to send the two brave hunters fleeing from the shaft stopping only long enough to grab their rifles. It was a quick trip back to their camp. One stood guard while the other packed up the camp, and they were headed for town before the sun went down.

MITCH'S COMMENTS:

The Matazel Wilderness Area is full of old mine shafts many of which have not been opened or inspected since the early 1900's. There is no hard evidence Bigfoot uses caves or shafts for shelter. However, early man used caves quite extensively. There have been caves located along the Little Colorado River showing signs of being used as a shelter. There was no indications the habitation was being used by a human, but the matting on the floor consisted of soft leaves, twigs and grass. Too much of the matting to be placed there by small animals such as a Pack Rat. Cats and bears do not make such mats. Recently there has been another set of caves discovered on the Mogollon Rim which contained the same type of matting.

CAVES ON THE MOGOLLON RIM

NOTICE THE MATTING AND BURROS

STORY 4

3 OCTOBER, 1974, TONTO CREEK NEAR MESCAL RIDGE ON THE NORTH SIDE OF HELL'S GATE WILDERNESS AREA.

"It's going to be a cold wet one," Greg complained as he pulled on his coat. "I hope the hike is worth it."

"It will be." Tom promised. "We get away from these over-fished holes, we'll get into some nice ones."

"I've never fished this side of the highway," Greg stated. "Always went the other direction towards the Fish Hatchery."

"Me too."

The two men stuffed their fishing gear and some snacks into their day packs and locked the vehicle. It was going to be a bit of a walk to get to a part of the Tonto Creek that very many do not fish.

The weather was not cooperating. It was over cast and looked like rain. The clouds were heavy and black. This did not deter the two fishers. From time to time a rain drop would catch the in the face. It was reminder as to what could come. But, weather around the Mogollon Rim was fickle. The clouds would come from the south and meet the high cliffs of the Rim. The clouds seemed to struggle getting over the rock cliffs, and this made them dark and angry. It could rain very hard, or not at all.

The men were prepared with rain gear, coats, food and water. They were both experienced outdoorsmen. They knew if it rained hard on the Rim there could be flooding in the steam below. But, they were willing to chance a cold wet night to catch some nice sized trout.

They decided not to waste the morning fishing down the stream, but to hike two or three miles down the canyon before trying any holes. They figured most fishermen would not go much past that point before turning around and heading back to civilization.

It was nine in the morning when they stopped at their first hole. They had been walking for a little more than two hours.

Tom was the first to score. It was a nice rainbow trout a little over nine inches in length. It was a keeper because they planned to have a trout lunch and keep fishing. They continued downstream alternating the fishing holes so each would have the turn being first to drop a line. They threw back anything under ten inches. It was a good day. The trout were hitting nicely, and the weather had been cooperating with no real amount of rain.

By lunch time they had worked their way down into the canyon where the sides were getting steeper and harder to negotiate. Any further down would require them to start getting wet. They would have to swim the next hole, and that defeated the purpose of all the rain gear. It was time to stop, have lunch and fish their way back to the car.

They found a good flat spot sheltered by the boulders and started a fire. Greg broke out some tinfoil, a lemon, and some salt and pepper. Meanwhile, Tom cleaned the four trout they had kept. As soon as the fire was down to coals, the tinfoil wrapped fish was ready for cooking. It was a glorious smell. They were very hungry and the smell of the cooking fish made them even more so.

Tom grabbed a stick and began to fish the foil packages out of the coals. He had retrieved one and set it on a rock to cool while he began to remove another from the coals without poking a hole in the tinfoil.

Suddenly both men were alerted to a loud splash in the stream below them. Both stopped what they were doing and moved down towards the stream. They walked down the stream for a few moments, and here was nothing in sight.

They turned around and walked back to the fire. Tom picked up his stick and started to resume the removal of the next fish packet. It didn't take long for him to realize he was short one fish.

"I know you are hungry, but you didn't have to take it." Tom accused Greg.

"Take what?"

"The fish. The one on the rock I had cooling."

"I didn't take the fish. Where would I put it?"

"Must have been a fox or skunk." Tom admitted.

"Yeh, and they were just waiting for the opportunity to snatch a fish."

Tom was successful in removing the three fish from the fire. The two men found comfortable rocks and sat down to share their three remaining fish. As they were finishing up, Greg noticed the first flake of snow.

"We got snow," Greg stated. We're going to have to hurry going back out."

"Well, I guess we had a pretty good day, but I don't want to be climbing over slick rocks trying to get back out of here."

The two men dawned their gear, and Tom doused the fire with water from the creek. There wasn't much chance of a forest fire, but he didn't want to take any chances. By the time they started back, the snow was coming down heavy. It was going to be one of those early storms that takes everyone by surprise. Snow in October.

It didn't take long before the snow began to stick. The rocks and boulders began to be very slippery and progress was slow. They had managed to travel about half the distance back to their vehicle, but it was getting late. Dark comes early in the canyons of the Rim and with the storm clouds, it was going to be pitch black soon. They had flashlights, but trying to walk out in the dark over the slippery rocks did not appeal to them. They were cold and needed to warm up.

Greg spotted a large spruce tree upstream and figured it to be a good place to stop for a rest. Its dense canopy made a perfect shelter from the snow. If need be, they could spend the night there and walk the rest of the way out in the morning.

Tom reached the tree first and wasted no time climbing underneath the limbs. He looked around at what could be their night's accommodations. It was large and roomy, nearly twelve feet across. He was able to stand upright.

Greg followed and was amazed how nice it seemed under the tree. It was dry, and there was no snow coming through the foliage. There was enough room to stand and walk around, even enough room for a small fire to keep it warm.

"Well, I guess we can stay here for the night." Greg announced.

"Ok, this doesn't look too bad." Greg admitted. "Guess we should get some fire wood together."

The two started looking around for dry wood. Neither of them wanted to go outside to look, but they both knew they would have to before it got completely dark.

"Hey," Greg was surprised at a white gleaming object near the trunk of the tree. "What's that?"

Tom turned around and bent down to get a better look. "Looks like a bone."

"What kind of a bone?" Greg questioned as he stepped closer to take a look. With his step forward, he heard something crack. It wasn't a branch, but It felt like one. "It's another bone."

"This one looks like a leg bone." Tom announced. "Maybe a deer?"

"this one looks like a rib," Greg commented.

"I think we found the resting place of some wounded deer. Probably got shot last season and climbed up under here to die."

"I don't think so," Greg spoke with a shake in his voice.

Tom instantly recognized the anomaly in Greg's voice. It was fear.

"What do you mean?" He was almost hesitant to ask.

"The deer had buddies." Greg pointed out a rock on the uphill side of the tree. On it sat four deer skulls. All neatly facing forwards towards them.

"Cat maybe?" Tom squeaked at first but cleared his voice. "Bear?"

"Never known either for housekeeping." Greg spoke in a hushed voice.

"So someone else found them and stacked them up for a sick joke." Tom was mostly trying to convince himself as he fished out his flashlight.

This slight action alerted Greg that Tom was getting worried. Tom pointed the light into the tree top. Right above them was a half-eaten elk skull. Right next to the skull was a wad of tinfoil with a partially eaten trout protruding from the side of the packet.

Suddenly the two men found renewed strength and energy. They did not stop until they had reached their vehicle and were inside the locked doors.

Comments from Mitch: This story seems to have some interesting aspects. The mention of a large spruce tree and carcass remains is quite likely true. We have located and documented a pine tree much like this in one of our search areas. The carcass was stripped of the meat, head was gone, and the front legs were missing. Predators such as lions, bears wolves and coyotes break and eat the bones. There was a large bed or nest made out of tree twigs and leaves. Witnesses were myself, Clay Randall and his wife Lorie Randall.

A NEST UNDER THE SPRUCE TREE

NEST AND CARCASS VIEW

CLAY AND LORIE CHECK OUT A FOOTPRINT
NEAR THE CARCASS. IT HAS FIVE TOES!

STORY 5

December, 2006 The Mogollon Rim Overlooking See Springs.

Lenny had been scouting for the upcoming deer season. It had been a fruitless day. He had not seen any game. There were plenty of old droppings, but nothing new or fresh. However, he did see one set of deer tracks on the trail he had used to get to this vantage point. He decided to walk along the Rim while using his binoculars to spot game or possible good hunting areas.

The view from the edge of the Mogollon Rim was amazing. It was almost as if the world lay at his feet. Vast acres of forest land lay before him. Somewhere out there was a deer just waiting to be put in his freezer.

He wanted and needed the meat. Sure he could make a living back in town, but he didn't like the idea of eating tainted meat. Over the years, he had read various articles of how the beef in the food supply was tainted with growth hormones, and steroids from the cattle ranchers, to fatten up and bulk the cows prior to slaughter. He had become convinced these hormones and steroids were being passed on to the populous through the meat. No wonder Americans were having such a hard time with obesity and health problems. One of the last articles he had read was about the possibility of these agents causing early puberty in the American children.

Lenny wanted the meat because he knew what deer and elk ate. Grass and bushes as nature had intended, not all the filler and crap the poor cows of the meat industry were being fed. He decided to go all natural, grow his own meat and vegetables. At least then he knew what he was getting.

Only problem was, he had no place to grow beef. He was going to have to hunt for his meat. Therefore, this hunting season was very important to him. He had managed to get drawn for deer hunt, and no telling how long it would be before he would be drawn for another.

`Arizona was growing too fast. Too many "transplants" as he called them. Of course, he was referring to non-native Arizonians. He hated the fact that he had grown up in Arizona, but someone who had only been here a year would have just as much chance to get drawn for a hunt as he did. To him, it was unjust. And, to top it off, the number of drawing applications verses the number of permits was vastly out of proportion. He figured he would be drawn once every seven to eight years and this time period was growing.

He wanted this meat to tide him over until he could make arrangements with someone to grow a cow or pig for him.

Now he had come to the edge of the world, and somewhere out there was a deer waiting for him.

His attention was suddenly snapped away from the Rim. There was a terrible ruckus going on not three hundred yards away from him. He could see the bushes moving and thrashing, and the terrible screams. Something was being torn apart in those bushes, and it frightened him. All he had with him was his lock-blade pocket knife. He didn't want any game rangers thinking he was out to pouch a deer.

The battle in the bushes raged on for what seemed to be a few minutes. And then it was over just as quickly as it had started. The bushes no longer moved. The screams had ceased.

His first thought was it was a bear that had cornered a deer, but the screams did not sound like a bear. He also thought it was a mountain lion, but that too didn't sound right. He had heard both before, and neither matched what was over in the bushes. His only problem was the area where the ruckus took place was long the path he needed to take to get off the point he was sitting on. There was only one way out without trying to go over the edge of the rim. He decided to wait for a while and then make his attempt to get back to his car.

The wind was not cooperating very much. It was coming up the rock face and blasting over the rim. It was impossible to hear any movement in the bushes. It also made it hard to discern any movement in the bushes caused by anything other than wind. He was hoping whatever it was would eat it's fill and then move on. Hopefully, it might catch his scent and decide to vacate instead of going on the hunt.

An hour passed, and Lenny worked up his courage to make his escape attempt. Quietly he slipped down the spine of the ridge towards the bushes. There was no noise or movement.

As he approached the area he could see an opening between the shrubs and large rocks. He couldn't resist taking a closer look He drew his knife and started down the narrow corridor. It opened up into a small area. In the middle of the area was a pile of stick and leaves. It was the kill.

Lenny didn't know why, but he was relieved to find that it was a deer. But, he was horrified to see that it had been literally torn apart. He had never known a kill to be this messed up. The legs were torn from their sockets and the bones were broken in several places. There were no knife marks, and there were no bite marks on the neck. The head had been ripped from the neck much like Lenny did to rabbits in the field. He normally just stepped on the head, grabbed the hind legs and pulled.

Surely a big cat would grab the deer by the throat and crush the air passages to smother its prey. But, remove the head? That usually happens as the predator is feeding. Not as a matter of preparation to eat. A big animal usually goes for the stomach first. The soft tissue is the easiest to devour in a hurry. The rest will keep for a later meal.

Later meal? This snapped Lenny back to reality. He was standing on an animal's prey. A very large and powerful animal. There was still only one way out. Lenny looked over the brush behind him. It was a sheer drop for forty feet. He looked at the pile of leaves and noted the absence of claw marks in the dirt. Instead, the marks were made like a four fingered rake with no claws. This spooked Lenny.

It was time to leave. Quickly he skirted down the corridor. His heart was pounding from the possibility of being ambushed just like the unfortunate deer. He could just picture himself about to clear the corridor when a large beast steps into the entrance. What would he do? Try to force his way through the thick brush? Turn and run to the clearing. Jump the cliff and hope he survives? Stand and fight with his puny pocket knife? What if a hairy arm reached out from the bushes and grabbed him?

Suddenly he burst out of the bushes onto the trail. He kept running. He had no idea if there was anything behind him, but he wasn't going to wait around to find out.

He could see his car. He was almost there. Almost safe.

He chanced a look back on his trail. Seeing nothing, he slowed down and walked backwards for a few steps. Then it grabbed him. He fell to his knees in fright. Then realized he was looking at a pair of slacks. It was a Forest Ranger.

MITCH'S COMMENTS:

I found this story to be quite informative in the possible methods a Bigfoot would use to hunt deer or elk. There have been reports of hunting methods to include running the game over a cliff, or ambushing the game in a very thickly grown area. I have even heard of Bigfoot using sharp sticks propped up along a trail in thick brush to impale any deer herded down the trail.

Recently, some ambush points were located on the Mogollon Rim. These points were constructed by using a large pine tree that was toppled over causing the root ball to stick in the air. The area underneath the root ball was hollowed out to create a large pit. The entrance to the pit was covered with limbs from the pine tree.

This ambush point was enhanced by creating a funnel of trees along the path which would serve to guide fleeing deer or elk past the ambush point.

I believe a Bigfoot would hide in the hole with the branches concealing the entrance. When a deer or elk got close enough, it would spring out and kill the animal with either a blow to the back bone, or impaling it with a sharpened object such as a leg bone from another kill. Or perhaps, the animal would be killed by a crushing blow to the head from a large stone.

THE CLIFFS OF THE MOGOLLON RIM

THE EDGE OF THE WORLD

THE ROOT BALL FOR THE AMBUSH POINT

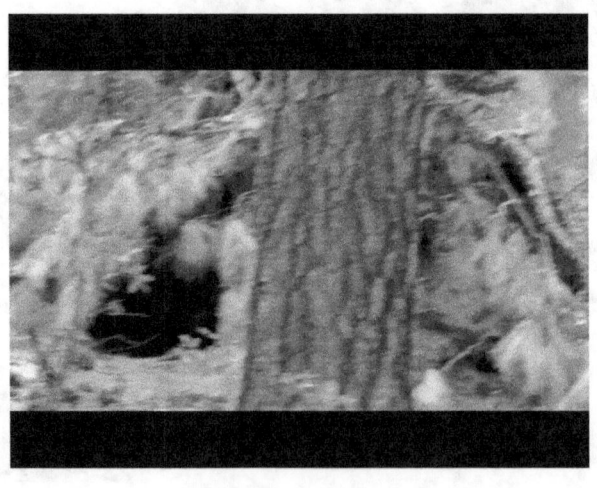

THE OPENING

**DAVID WAITE INSIDE THE ROOTBALL
AMBUSH POINT**

AMBUSH POINT

THE TRAP DOOR IS CLOSED

CAUGHT ON A GAME CAMERA, THIS ELK
APPOACHES THE AMBUSH POINT. NOTICE
THE HOLE IS COVERED. COULD THE TRAP BE
SET WITH A BIGFOOT INSIDE READY TO
SPRING OUT? WE CHECKED THE TRACKS AND
FOUND THERE HAD BEEN A STRUGGLE WITH
PRINTS PUSHED UP IN THE PINE NEEDLES.

STORY 6

17 FEBRUARY, 2007
CHEVELON RETREAT, MCALLISTER AND MAIN STREET

It was a beautiful day in the early spring. The weather was warm for the time of year. There were still patches of snow in the shadows of the junipers, but the roads were mostly solid and not muddy.

Mike was starting his trips to his remote property to check his apple trees and to ready the drip system for summer use. In the previous years, Mike had built a water tower to provide water pressure for the drip system. He had installed battery operated timer on the system, and the tank held 250 gallons which would last nearly a month before needing a refill.

Two years before, he had acquired a large camper shell to stay in while he was at his property. It would sleep four people, but usually he only had himself and two grandsons. On this trip, Mike was by himself.

Mike pulled off highway 260 near Heber's Circle K on the west side of town and headed north on road 504. This was going to be a quick trip to check the status of his property, check the water level and his camper. On previous trips he had found his camper sitting upside down off the wood pallets he had laid down for a foundation.

The first time he found his camper upside down, he thought it might be the wind. However, he had second thoughts when he tried to upright the shell. He could not move the camper on his own. He ended up hooking his winch on the front of the truck and pulling it over upright. It took some doing, but he was able to center the shell back on the pallet. This was done by repositioning the truck several times and using the winch to slide it into place.

The second time he found the camper on its roof he believed it to be an act of vandalism. He repeated the steps of getting the camper back on the pallets. However, this time he had brought some two by fours to make sure it wasn't the wind. He had nailed them to the sides of the camper making the two by fours into support legs.

On this trip, Mike was going to try to solve the mystery of why his camper was being turned upside down. He had brought a thermal sensing game camera. He was going to install it on his water tower overlooking the camper. He was a little hesitant to do this because the vandals may spot the camera and take it. However, he wanted to know.

Mike pulled up to the turn off for road 153 which would take him to the southern side of Chevelon Retreat. He drove past the Potato Wash burn. It was sad to see all of the burnt trees and barren land caused by the forest fire. There were acres and acres of scorched land and burnt trees. Before the fire, Mike had usually spotted elk or deer, but there were none since the fire.

Mike remembered how the big concern of most of the valley people in Phoenix was the fire reaching the high tension power lines just south of the retreat. There didn't seem to be much concern over the homes from the valley people, but the fire fighters had put up a valiant fight and were able to contain the fire before it reached the homes in the Retreat.

At the power lines, Mike turned east on road 95 and headed for the Squaw Tanks. He noticed how the fire fighters had cleared the trees and brush away from the road and power lines to make a fire break. There had been a lot of heavy equipment in the area to establish the fire break. They had done a good job.

When Mike reached the tanks, he turned north and crossed out of the last of the National Forest land and into private land. He passed Larsen's Tank and noticed it was empty even though there was some snow still under the juniper trees. It was going to be a long, hot, dry summer. The prairie grass was already dry and yellow. Too much of it too. Somehow he was going to have to get it cut down so his property wouldn't be at risk.

Finally, he was on Main street and approaching MacCallister. He thought how funny this sounded. Main Street of nowhere. His nearest neighbor was a quarter mile away, but that was why he liked his place. He could go there and see or hear no one.

Mike had plans for this land. He was going to eventually build a cabin. He had already planted a dozen apple trees, established a drip system, and started putting up a fence line. He had managed to put in one hundred and ten fence posts, but that only covered two sides of the property. He had some wire to string, but that would have to wait. He was going to have to get the drip system ready for operation. He had drained it for the winter, and now he needed to check for leaks, and fixes prior to filling the water tank.

Mike pulled onto his property and was relieved to find his camper was still standing upright. It had not been rolled over since he nailed the two-by-four legs to the camper. Time was short, so he went right to work on the drip system. It seemed to be in very good condition. A few small fixes, and it was ready to go.

He climbed the ladder to the top of the tower and placed his game camera looking down on the camper below. He determined where the edge of the sensing would be and climbed own off the tower. He took one of his shovels and stuck it in the ground as an indicator to him so he wouldn't accidentally set it off.

After hitching up his water trailer, Mike departed to Well number four. It took him about an hour to travel the two miles and fill the water tank. Going was slow due to the roughness of the roads. But, he soon returned and climbed the water tower pace the pump hose in the top of the tank. He noticed his game camera had taken a picture.

At first he thought it was his neighbor's dog. The animal was small and black about the size of a small Labrador retriever. But the neighbor's dog wasn't black. Then he figured it was a bear cub. But there was no snout. Then he noticed the feet. The rear feet were large, and the front feet looked like hands.

MITCH'S COMMENTS: It is peculiar that the camper was turned over onto its top several times. The camper is way too heavy for a single person to move. It would take a very strong microburst to budge the camper off of the wooden pallets. This could happen once, but two times?

There is a large population of pack rats in this part of the country. The owner of the camper found that he could leave a five gallon bucket half full of water on the step of the camper. He would leave a two-by-four leaning on the bucket edge to the step. Each morning he did this, he would have three to five drowned rats in the bucket.

It is possible the rats were a food source, and the pallets under the camper created the perfect rat habitat. Perhaps, the camper was being overturned to allow access to the rats in the pallets?

THE CAMPER AND WATER TOWER

THE CAMPER WITH LEGS NAILED IN PLACE AND
STRANGE SHADOW IN FOREGROUND

STORY 7

3 March, 2007 Vernon, Arizona.

John William was nearly asleep in front of his TV when he heard his wife screaming for him. Her voice came from outside, and he heard her slam the back door to the house. She sounded nearly hysterical.

He snapped to consciousness and came to her assistance. "What is it?"

"It's a big bear!" She was shaking with fear. "It's by the garbage cans."

John grabbed his twenty two magnum pistol and stepped out the door. He knew the gun would not kill the bear, but he was hoping to scare it off.; maybe enough of a scare to make it afraid of humans.

He slowly walked out towards the cans with all his attention on any movement in the bushes behind the cans. But, the animal was not there. It wouldn't be the first time a bear had tried to get into the cans. The Johnsons had learned a long time ago that ordinary cans were not adequate for keeping trash invaders like bears out of the garbage. Too many times they had to clean up the litter and mess left behind. The new bear proof garbage cans seemed to do the trick. However, it did not stop the occasional uneducated bear from attempting to gain a free meal. John was going to educate the bear.

There was a small ravine that went down his property line paralleling his drive way and property line. This ravine passed just behind the garbage cans and back along the south side of the house into the woods. This is where John thought the bear would be. He studied the ravine very closely on his way back to the house.

Suddenly, he heard a low grunt. It was more of a huff, and it didn't sound like a bear. The sound came from the opposite direction from what he had anticipated. He slowly turned and looked over his left shoulder. There was a very large figure standing in the drive way not more than ten feet away.

Knowing that his pistol would prove useless in an all-out confrontation with a bear that large, he decided to try to bluff his way out. He instantly puffed his shoulders up and made himself look as big as he could as he turned around to face the animal.

To his surprise, it was not a bear. It was a very large hairy man that stood a good foot taller than John's six foot frame. It was standing upright and covered in dark brown hair. There was no clothing.

The hairy man seemed a bit surprised that John stood his ground, and tilted its head to one side as if it was wondering about something.

While this was going on, John had decided not to shoot the man, if it was a man, unless it attacked. He shuttered at the idea of finding out that it was some poor, lost soul that had been lost in the woods for a couple of years.

The two stood facing each other. The hairy man looked at the pistol in John's hand, and John could swear he could see a look of concern in its face. All the time the man-creature was rocking slowly back and forth with its head tilting in the direction followed by the shift of body. It seemed to be rhythmic and non-threatening gesture.

John put the pistol behind his back but was still ready to use it. As soon as he had done this, the hairy man turned slowly and walked away to disappear into the bushes.

John stood in his drive way wondering about the strange incident. John figured the slow rocking motion was an indication that the man-beast wanted to leave peacefully, but was not about to back down from a fight. When John had put the gun behind him, it was giving the creature permission to leave. John now believes both just received a crash course in communications, and they understood each other.

MITCH'S COMMENTS:

Most Bigfoot researchers will agree the biped has methods of communications including short range vocalizations such as speech, and non-verbal (body language) to supplement the verbal language. Most humans have lost much of the non-verbal skills, but we do use hand gestures.

Then there are the methods of communications used for long distance, location, and hunting. These communications include bellowing, moans, whistles, animal/bird mimicking, rock clacking, and tree knocking.

I believe the most universal language is the non-verbal means of communications. It is pretty straight forward, conveys feelings, and is very hard to lie.

The swaying back and forth from side to side seems to indicate a decision has to be made or there is some sort of question to be answered. In the story above, it appears as if the Bigfoot was asking if he could leave peacefully without getting a bullet in the back. It appears, John read this communication correctly and gave the proper non-verbal response by placing the gun behind his back indicating there was no desire a confrontation. Not a word was spoken, but both parties knew what the other meant.

It is this story that started me thinking about Bigfoot Protocol. How should you act in a nose to nose encounter and how the improper response could lead to a confrontation. I started developing techniques to be used in a Bigfoot encounter:

THESE TECHNIQUES MAY SAVE YOUR LIFE.

Keep your voices low.

Do not smile showing your teeth.

Do not look them straight in the eye.

If you encounter an adult male, do not stand upright, keep your head low. You can look at them, but only short glances with your head low. A slow rocking motion from side to side is asking for permission to leave. Clearing your throat in a low, gruff "ah-hmmm" manner announces you mean no aggression.

Do not reach out to touch them

If charged, do not run. Do not show fear. If attacked, fall to the fetal position and ball up the best you can.

Do not flash them with bright flashlights or white lights.

Do not chase them.

Do not show your gun. They know what it is, and the sight of it may cause them to act violently.

Do not make quick or sudden movements.

Make no threatening gestures. Drop any walking sticks or anything that could be a weapon.

If you hear them in the bushes, do not act like a "deer in the headlights" or stare into the bushes trying to see them. Go about your business, but keep close watch with your eyes, not your head.

STORY 8

18 MARCH, 2007
Just off county road (CR4131) off Highway 180 on top of Picnic Hill South East of Springerville, Arizona.

"It's been a long day," Rich remarked to his son. "We can stay at my sister's house if you want to."

Dave looked at his dad and smiled. "I want to camp. Let's make a fire and cook some Dutch oven biscuits."

"Ok," Rich agreed. "We'll have to grab some wood while it's light and put the tent up in the dark."

"That's fine."

The two had been traveling all day on their way home back from Neches, Texas. Dave had bought some land, and his dad had accompanied him on the trip to inspect the property. It had been a whirl-wind, four day trip from Mesa to the far side of Texas and back.

For the return, they decided not to stick to the interstate, but to go through Roswell, New Mexico to see the Alien Museum. From Roswell, they would travel to Springerville, Arizona to stay with relatives. From there it would be an easy four hours to home in Mesa the next day.

They were running out of daylight when they came over the ridge overlooking Round Valley. Instead of proceeding into the town, Rich turned his truck onto Highway 180 and headed up Picnic Hill to the National Forest. They were in a hurry to find a camp spot, gather some fire wood, and pitch camp before it got too dark.

"You know this hill has a couple of stories to it," Rich commented.

"Ok,"

"This is the hill here my mom talked my dad into proposing to her. According to my dad, they were up on top of the hill for a picnic, and were headed back home. As they started down the hill in their car, mom pulled a knife and poked my dad in the ribs. He said she made him promise to marry her or they were going to both die. So he proposed to her and married her a short time later."

"Grandma? No way."

"Well that's what my dad always claimed. Of course, my mom always had another story about that. But, her version is much more boring."

"Ok, so what is the other story?" Dave questioned.

"Back in the early 1900's the Mogollon Monster was heard screaming on this hill. The entire town thought it was a woman being tortured or kidnapped. They formed a posse and started chasing it. Finally, they figured it out. The Mogollon Monster was leading them towards Escudilla Mountain. Later, they burned Escudilla to try to get rid of the pesky beast."

"Another family tall tale?" Dave smirked a little.

"Laugh if you want, it happened."

The truck topped out on Picnic Hill, and they slowed down to look for a road to possible camp site. It didn't take long. The dirt road took them across the top of the flat hill and started down a slope. At that point, they found a small dirt road headed west. It led them to a camp area, which was surrounded by pinion pine trees.

The two went to work immediately gathering wood from under the trees. It was still light enough to see, but the sun had been down for some time. They were lucky and found some wood that was dumped by someone from town that didn't want to take their garbage to the landfill. They decided to use the wood from the pine trees for cooking and then the wood from the dump pile for light and heat after the cooking was done.

Dave went about pitching the tent while Rich stoked the fire and prepared tinfoil dinners. It was at this point Rich noticed a strange sound.

"Dave."

"What?"

"Do you hear that?"

The two listened, but the only sound was the slight breeze rustling through the pine needles of the surrounding trees.

"What was it?" Dave queried.

"It sounded like a tapping noise. Kind of rock on rock, but a long ways away."

"Nope. Didn't hear it."

"It stopped. I don't hear it either."

The two went back to their activities. Dave finished the tent and rolled out the air mattresses.

"OH, crap!" Dave was obviously not happy.

"What's up?"

"My batteries are dead on my mattress inflator."

"Well, you'll have to do it the old fashioned way." Rich teased. "Better get started on it. And, I like mine extra firm."

Dave mumbled something, but could be heard huffing breath into one of the mattresses. Rich finished up cutting the vegetables for the tin foil dinners and wrapped them tightly in tin foil pouches. Taking the shovel, he stirred the fire a bit and retrieved some of the hot coals and spread them out for cooking. On these hot coals he placed the tinfoil packets and then carefully shoveled more hot coals onto the top of the dinners. Next he started to stoke the fire.

He stopped as soon as he heard it. The same clicking noise, but this time it was accompanied with some low noises. Almost sounded like words, but had a moan to it.

"There it is again," Rich spoke quietly to Dave.

"What?" Dave was still inside of the tent puffing on the second air mattress. "I don't hear anything. I'm a little dizzy, but I don't hear anything."

"It's still a long ways away." Rich walked around the truck into the dark to look the country side over without the firelight overpowering the starlight.

"I can see another campfire," Rich commented. "It's a long ways away."

"Could the noise be coming from there?"

"I don't know. The fire is a good mile or two away. It's definitely on the other side of the highway. Maybe if the wind carried the sound."

Rich went back to the fire and stoked it with more pine tree wood.

"The coals will be about right for your biscuits."

Dave finished up the mattress and spread out the bedrolls. Finally, he crawled out the tent and came over to the fire.

The Bisquick is by the ice chest." Rich instructed. "Milk is inside. You can use the big frying pan to mix up the dough."

Rich retrieved more coals from the fire and made a bed for the Dutch oven. Dave mixed up the dough and retrieved the butter and raspberry jam from the ice chest. Soon the biscuits were cooking.

"The dinners will be ready in about five more minutes," Rich announced. "How long for the biscuits?"

"About ten minutes on the coals, then we'll have to let them brown on top. I have a lot of dough left over. Maybe a second batch. Can you eat that many?"

"No way."

"What shall I do with the extra dough?" Dave pointed to the large frying pan.

"Just leave it in the pan. We can maybe use some of it in the morning to make pancakes."

"Ok. What about bears?"

"I think we are Ok. Just leave it in the back of the truck."

After a short while the two had their feast. The tinfoil dinners were done to perfection, and were totally consumed. The biscuits were golden brown and fresh. The butter melted into them.

"There's nothing like hot, fresh Dutch oven biscuits with butter and raspberry jam." Rich commented as he jammed his third one into his mouth.

Dave was busy dressing his next biscuit with toppings while he washed down the first one with a glass of cold milk.

"I'm glad we camped," Dave mumbled finishing up the biscuit. "This is great."

"And," Rich added. "There are no dishes to do. Just wipe out the oven and we're good to go."

The two sat round the campfire talking and poking the fire with sticks. Eventually all the wood was consumed. They even burnt some of the wood from the trash. The conversation slowed, and eventually both were snoozing in the camp chairs.

"Dave." Rich reached over and woke up Dave. "It's time to hit the sack."

"Ok."

"I'll put out the fire."

It was 11:30 pm when they climbed into their bedrolls.

"You hear that?" Rich quietly spoke to Dave.

"What? What time is it?"

Rich flipped open his cell phone and looked at the time. "It's 1:30am. Something's outside."

"Probably a skunk or raccoon." Dave groaned and rolled over to go back to sleep.

"No, that was my frying pan. You did leave it in the back of the truck?

"Ah, yes."

"Sounded like it got dropped in the truck."

Rich located his pistol and flashlight and he quietly slipped out of his bedroll. Slowly and as quiet as possible, he unzipped the flap of the tent to peer outside. It was pitch black. Only the starts above could be seen. The trees shadowed everything else.

If it was a bear, Rich was prepared to try to scare it away. He was hoping to give it enough of a scare that it would not like being is such close proximity to a human anymore.

Rich readied his flashlight and pistol. He aimed his flashlight at the truck and turned it on. There was no movement. Nothing was there. Rich stuck his head out of the tent and inspected closer. Nothing seemed to be around the truck. Thinking the bear might be on the dark side of the truck, he flashed his light low and looked under the truck for feet on the other side. Nothing. Whatever it was—was now gone.

"It must be a skunk or raccoon. Rich stated as he climbed back into the tent and zipped it closed. "There's nothing out there now."

Dave just mumbled and went back to snoring.

Rich popped the clip of his nine millimeter and ejected the shell onto his bedroll. He checked the chamber twice before releasing the slide and un-cocking the hammer. He slid he unused round back into the magazine and installed the magazine back into the pistol. He now had a full clip, with nothing in the chamber.

A few moments later and Rich was back in his bedroll and thinking about his frying pan. He figured if it was a bear, it would eat the dough and be back later to raid the ice chests and to look for scraps. Perhaps, he should try to sleep light. That was his last conscious thought as he closed his eyes and drifted off.

Suddenly, Rich's bed lifted up and rolled spilling Rich out of his bedroll onto the tent floor. There was a funny metal clank that sounded like his aluminum frying pan hitting something hard. Rich found his pistol and shucked a shell into the chamber. "I'm awake now!" He told himself out loud.

Dave just groaned and continued to sleep.

Rich looked at his phone to check the time. It was 4:30am.

It was quiet. The tent canvas moved slightly in a breeze. There were no sounds other than Dave starting to wake up. Rich strained his ears to listen for footsteps or activity outside the tent. There was nothing.

"Did you feel that?" Rich asked Dave.

"Feel what?"

"I just got rolled out of my bed, and you didn't feel it?"

"No. Just you moving around."

"Something hit the tent on my side. It bounced me off the air mattress!"

"You sure?"

"I'm not sleep walking here in the middle of the tent!"

Rich unzipped the tent and peered outside. He couldn't see much without the flashlight. He fumbled around for a few minutes and found the light. He fought the impulse to turn it on inside the tent. He thought this might not be a good thing if something was standing outside and could see shadows from inside the tent.

He stuck his hand and light outside the door flap and turned it on. The area in front of the tent lit up. There was nothing in the camp area

Rich moved cautiously out of the tent while constantly checking both sides of the tent. Nothing. He didn't want to, but he knew he had to check out the area behind the tent. It was dark and the flashlight was flickering and growing dim. What a time for the batteries to start going dead!

He quietly moved down the side of the tent and peered around the back corner. He let out a sigh of relief when he discovered nothing. "From now on, I carry a propane lantern." he thought to himself. "No more camping in the dark."

Returning to the tent, Rich zipped up the flaps and climbed back into the bedroll on the air mattress.

The rest of the night Rich contemplated the possibilities. Was it a gust of wind that almost collapsed the tent? Or was it something more? The wind had been very light and barely moved the tent sides. It was a calm night. Could it have been a bear? Not likely. It wouldn't be so quiet. A bear would have gone after the food in the back of the truck. Litter would be scattered everywhere if it was a bear. So what threw him out of bed and why? Was it his snoring? He was famous for his snoring.

Finally, it was light enough to see. It was time to find some answers. A close inspection of the area netted no footprints. With the exception of the circle around the campfire, the area was covered in pine needles and short crops of grass. There was no footprint evidence of anything being in the camp.

Dave finally climbed out of the tent and noticed Rich studying the ground. "Loose something?"

"Not much, just sleep."

"I was out like a light." Dave commented.

"Yes, I know. You could have been eaten by a bear and not even woke up for it."

"Ok, so what happened?"

"I got dumped out of my bed last night."

"You fall out?"

"No. Something nearly picked up on my side of the tent and rolled me out of bed."

"Gust of wind?"

"I don't think so. You did leave the frying pan with the dough in the back of the truck? Right?"

"Yes."

"So what is it doing in the tree?"

Rich pointed out the frying pan about eight feet up a ponderosa pine tree next to the tent. The pan was nestled in the limbs. Rich retrieved a stick and knocked the pan out of the tree. The sticky biscuit dough was gone. The pan was completely clean.

"Well," Rich laughed. "Something liked your biscuit dough."

Dave walked over to the side of the tent, and called Rich over to look at something.

"I drove all the tent pegs within an inch of the ground," Dave explained. "Except this one. There was a rock in the way. The tent peg was sticking up about six inches. It looks like something tripped over the peg and nearly yanked it out. Look, it landed alongside the tent. All the needles are pushed forward."

"You mean it nearly landed on me. That means when it landed on the edge of the air mattress it launched me into the middle of the tent."

"And, it probably wasn't a bear either," Dave added. "When on all fours, bears don't trip like that. Whatever it was did a face plant into the dirt. That's probably how your pan got into the tree."

"I wonder," Rich smiled as he mused over the situation.

"What?" Dave replied.

"I wonder if it got embarrassed when it tripped over its own big foot."

MITCH'S COMMENTS:

Although there was no Bigfoot sighting in this story, we felt it qualified due to the peculiar circumstances.

The frying pan moved from the back of the truck to a place high up in a pine tree. Could a raccoon do this? Possibly a bear might be able to take it up the tree and lick the dough from the frying pan.

Would a bear trip? Not likely unless it was waking on its hind legs only. That's the advantage a quadruped has over a biped. They don't trip as easily. A bear would certainly be heavy enough to launch Rich off of his air mattress into the middle of the tent. But, can a bear clack rocks?

THE FRY PAN

STORY 9

IT EXISTS!

It was fiercely cold. It bit at exposed flesh leaving it raw and numb. It was a mistake to take a deep breath. The nose hairs in froze stiff and acted like tiny sharp knives when he touched his nose. It was going to be a long, cold bitter night.

Marshall had realized he was in trouble when he saw the clouds move in and butt up against the Mogollon Rim. They seemed to strain at the cliffs trying to move up over the rocky spires and escape to the north. But they were too heavy. They had to drop the snow.

The day had started out nice enough. It was unusually warm for the time of the year, and Marshal had decided to go look for Bigfoot. He wanted badly to find some kind of evidence to show that the Bigfoot existed in Arizona.

He and been teased by the on-line Oregon and Washington researchers claiming Arizona was not a habitat in which Bigfoot could exist. To them, Bigfoot could only exist in the deep old growth forests of their states or Canada. The idea of Bigfoot being in Arizona was laughable.

He thought about how absurd it was that they pooh-poohed his ideas and demanded proof of the existence of the Mogollon Monster when they had not even proven that Bigfoot really existed at all. Like their Bigfoot really existed at all. He thought is was funny how the rest of the world looked upon them as being one of "those" people. Now they were doing it to him.

Marshal knew he'd probably find nothing. That was the usual case. But, he was going to keep looking for proof. He thought about that as he trudged through the snow.

Proof. What would that be? A photo would be written off as being "photo-shopped" and declared a hoax regardless of its authenticity. A footprint? That too would be declared a hoax. He would be accused of stomping around in wooden feet. An audio tape wouldn't hold much weight either, and video was becoming less believable with all the special effects that the computer had made possible. After all, look what they had done with the Patterson footage. No one knew for sure if it was real or a hoax.

Marshal was prepared. He figured if he could get a photo on thirty five millimeter film and on video he might have a chance. He had. his Bigfoot kit together. He had a video camera, his thirty five millimeter camera, and a disposable camera. His kit also had Fix It plaster for casting footprints, and he had a tape recorder just in case there was some vocalizations. He had brought warm clothing, and had comfortable hiking boots. He also had a coat, a bottle of water, and some snacks in his backpack.

All of this, and he was not prepared for the weather to do this. It snapped cold and was snowing like he had never seen before. The snow wasn't coming down in flakes. It was more like dollops of whipped cream. The ground was warm and resisted the snow sticking for a while, but was quickly overwhelmed. It was going to get deep quick, and Marshal was at least a two miles from his car. Visibility was becoming non-existent.

Suddenly, the snow stopped. Not a single flake fell. It was un-natural. It was almost as if God had turned off the water faucet to his snow maker leaving about a foot of white stuff on the ground.

Then he heard it. It was a deep wobbling sound followed by a couple of clicks of the tongue. It couldn't be a turkey. He had heard wild turkeys before. Quickly he scrambled for his equipment.

Video tape? Audio tape? Which should he use? The Audio tape would pick the sound better, but what if the source stepped into the open? The Video tape may not pick up the sound as well as the tape recorder.

Both, he would use both. He grabbed the tape recorder hit the record button and tossed it on the backpack. Snatching up the video camera, he hit record and pointed it in the direction of the sound. Nothing. Not a thing happened. No further sound and nothing stepped out from behind the bushes.

Marshal sat down in the snow with his camera still going. Perhaps, he was presenting too large of a profile for the creature to show itself. Still nothing happened, and all was quiet.

Marshal continued to tape for fifteen minutes, and finally decided that whatever it was had moved on. Whatever it was, he had missed his chance. He shut off his video camera, and reached down to rewind his tape recorder.

As soon as the tape recorder clicked off, it happened. The same noise and a quick whistle. Marshal flicked his recorder back on and raised his camera just in time to catch a snowball square in the face. It hurt, and stung his already cold cheeks and ears.

He wiped his face clean from the snow, and noticed the only sound was him cussing under his breath. His tape recorder had been knocked into the snow, and he had to go fishing for it. As he found it, he caught a flash of a shadow moving through the bushes and he whipped his video camera up to catch the movement. But he was too late. It had disappeared.

Then it came to him. This was no Mogollon Monster. It was someone playing tricks on him. The snowball was the give-a-way. Someone was out there in the bushes laughing themselves silly. Bigfoot had never been reported to have thrown snowballs. Rocks, yes, but not snowballs. This angered Marshal.

"OK," he called out in an irritated tone. "That was funny until you put someone's eye out!"

He couldn't believe he said that. He sounded like his mom when one of her kids ran to her crying because play was a little too rough.

Marshal turned off his video camera and stopped the recorder and placed them back into his bag. Again, the noise started low but ended in more a whine than whistle.

"Very funny!" Marshal spoke out loud as he swung his pack over his shoulder and headed in the direction of his car.

His reply was answered by another noise and a whistle just like the first and it was followed by another snow ball. It missed.

"That's it!" Marshal shouted. "I'm going to kick your butt!"

Marshal dropped his pack and grabbed at the snow making a large snow ball. He threw it at the last sound in the bushes. He heard another whistle and it was followed by another snow ball. Marshal returned fire. Soon it was all out war!

Marshal found himself packing snow balls and throwing them as fast as he could. He was really working at it, but having fun.

Suddenly, another noise sounded to Marshal's right. It was a powerful snort followed by a couple of Huffs. The snowballs immediately ceased. Marshal saw a dark figure a little smaller than himself moving though the bushes in the direction of the older more powerful sounds. He had the feeling his opponent had just been called home.

"I had you!" He called out as he picked up his pack and started back towards his car.

He thought about it as he walked. He thought it was a bit strange that a kid would be out there camping in this kind of weather. And, it was strange how not a word was used on his opponent's part. For that matter, when he was called home, there were no names called by the parent. Just an irritated snort and huff, that clearly got the point across. The games were over.

Could it be? Could he have just been in a snowball fight with a Mogollon Monster? Nah, it couldn't be.

Marshal trudged through the snow on his way back to the car. About a mile out, he found two sets of foot prints cutting across in front of him. One set of prints were huge. At first he thought they were made by snow shoes. The second set was much smaller, and about a size ten. Upon close inspection, Marshal could see toe prints at the bottom of the print hole. Both sets were barefoot.

Marshal sat down in the snow dumbfounded. Here was his proof. He had been in a snowball fight with a Bigfoot, and no one was going to believe him. He had no evidence on his cameras. In fact, he realized the whole thing was a game with is opponent. It had been playing with him from the beginning.

He thought about casting the prints, and photographing the evidence, but then he thought better of it. Why go through all the trouble to convince no one. He had all the proof he needed for himself, but he had no proof at all. Let North West researchers and others chase their Bigfoots into oblivion. He would not do that, and the less they knew of Arizona, the better.

MITCH'S COMMENTS:

When I read this story, I thought to myself, "This is me!" I know exactly just how this guy feels. However, I have never been in a snowball fight with Bigfoot.

I have run into resistance from researchers outside of Arizona, but most of them think of Arizona as being a dry desolate desert. They can't imagine the lush Ponderosa Pine trees that thrive along the Mogollon Rim. This stand of trees is the world's largest growth of Ponderosa Pines, It extends from West of Flagstaff all the way across the State into the State of New Mexico.

It is funny that Bigfoot Researchers of the Northwest, Canada, and Alaska won't believe a Bigfoot might be in Arizona. The last Bigfoot researcher from Alaska I talked to stated, "You have no proof, where is your proof such a creature could exist in Arizona."

I answered. "I have the same proof you do, but you. I have Squatch-blob photos, videos, hair, footprints, vocalizations, and multiple eye witness encounters."

I would expect that kind of comment from a unbelieving skeptic, but to get it from someone who believes? But, won't believe the same evidence he presents just because it comes from Arizona?

PART 2: MORE MOGOLLON MONSTER FIELD RESEARCH BY MITCH WAITE

Memorial Weekend, 2008 by Susan Farnsworth and Desert Rat. We decided to check out a well and spring that is about ¼ mile above the Falls. We worked the area looking for tracks and any sign of the Mogollon Monster. It was only going to be a quick over-night stay, and we were going to return on Memorial day.

As usual, we saw nothing around the spring or well. Of course, and elephant could have walked through that area and not be noticed because of all the pine needles on the ground from the very dense growth. The day was winding down and we decided to find a campsite in a small clearing. We were not sure of the fire restrictions, so we decided not to start a fire. Especially, since we were very near the Forest Service lookout tower. It was getting late and we didn't want to look for wood in the dark.

I got busy with dinner on the propane stove, while DR set up the dome tent and air matrices. It was a large 6 man tent with the flexible rods running up over the top to create the dome. The rain fly was installed over the top when the dome art was up and the tent was staked out. All-in-all, a pretty stable tent.

After the meal and cleanup of dinner, we decided to turn in early. I decided to for-go the usually ritual of placing my motion detectors, game camera, and voice activated taper recorder on the edge of camp. In fact, I left them locked in the truck along with my hand-held camera. But, I was tired and figured it was safe because we had combed the area pretty well during the day. Nothing was out there. (DOH-- a real not so smart thing to do. I'm supposed to be the expert (definition of an expert: A drip under pressure).

About 2:30am I awoke to a very foul stench. It smelled like decomposing fish. It was very strong and was turning my stomach. I was laying on my back facing up when I opened my eyes. Instantly, I noticed the tent side was about two inches from my nose. I was terrified. Something was actually pushing my side of the tent down on top of me.
I slid my hand out of the bedroll and located my pistol while quietly calling DR to wake up. Suddenly, the tent flipped back up into the dome position. If course, I screamed my lungs out, and rolled to the center of the tent.

I never saw a man move so fast as DR did. He was out of the bed and had pistol in hand almost before I reached the center of the tent. He started making very angry noises and hitting the sides of the tent with his pillow. I calmed down enough to turn on an electric lantern, and threw it outside. This allowed us to see under the rain fly a short distance outside, but hopefully nothing outside could see our shadows from the inside. Nothing was near the tent. I must admit, if anything would have touched the tent at that point, we would have blasted it, and the tent would have had a new door!

We waited for a few moments (which seemed like an eternity to me), then DR slipped out side. He picked up the lantern and did a quick walk around. Me? I headed straight for the truck. Once inside, I started the engine and turned on the lights. I was ready to hit the gas if anything was spotted, all the while having the door open for DR to jump in.

We decided to go to the nearest town and come back in the daylight to get our tent and stuff. We found no foot prints, but there were some places where the pine needles had been pushed back on my side of the tent.

I still quake in my boots whenever I think about it. However, I do kick myself for not setting my motion detectors, game cameras, and recorder. And, if we would have seen the creature, I would not have taken any pictures. That was the last thing on my mind. I would have been too busy beating a hasty retreat! I also know that I sound just about like the woman in the horror film when they get surprised by the monster.

I wonder why my side? Was it curious? Or Hungry?

Two weeks later an expedition of four (Kyle Barentine, Mitch Waite, Preston Smith, and Alex Hearn) returned to the tent site to investigate the event. While searching the area, they found many impressions that seemed to be made by extremely heavy and large feet. A couple of bedding areas were located where branches had been piled up under the low brows of a tree to make a bed. There were broken branches with leaves laying all around the area.

KYLE BARENTINE

MITCH WAITE

PRESTON SMITH

ALEX HEARN

As the group made their way back to the road, they ran into a Forest Ranger, and they stopped to talk for a while. The Forest Ranger (Dave) stated there had been no bear in the area for two years, but he had been finding scat that he could not identify. He had not thought much about it until he ran into us. While we were talking to the Ranger, Kyle located our first major find.

As the Ranger drove away, Kyle called us over to the side of the road. We were dumb stuck with what he had found. It was a footprint--19 inches long, 9 inches through the ball and 4 inches wide in the heel. Alex provided the plaster to cast the print. We determined that this area needed further investigation.

FOOTPRINT BEING REMOVED FROM THE DIRT

FOOTPRINT CAST PLACED ON TOTE LID TO KEEP IT BROM BREAKING

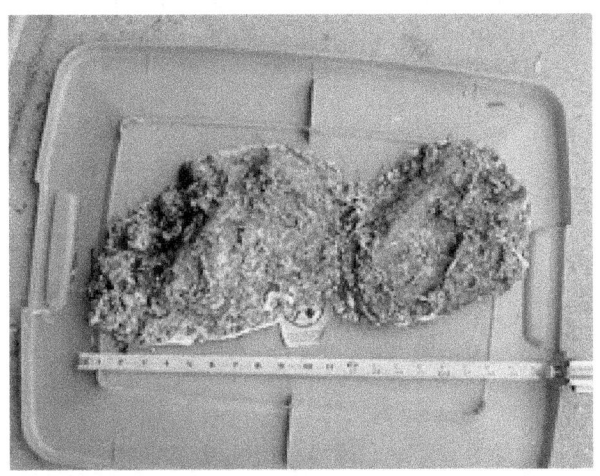

19 INCH FOOTPRINT CLEANED UP

EXPEDITION 1, 6-7 June, 2008 (Friday-Saturday)
by Mitch Waite

Participants: Kyle Barentine, Preston Smith, Alex Hearn, Mitchell Waite

Purpose of the trip: Two weeks earlier, Mitch and Susan went camping in the Mogollon Rim area. During the middle of the night, the dome tent was squashed down on Susan's side and held about two inches above her nose. The tent was released and popped back up to the dome shaped. During this time there was a very foul odor which Susan believes was that of the Mogollon Monster (Arizona's Bigfoot). Both made an evacuation of the tent and made it to their vehicle, but saw no animals. They left the area to return during the day to pack up their belongings.

Two weeks later this expedition was organized to go back and try to find out what had happened.

Departure and arrival at camp: Preston, Kyle, and myself departed Mesa, Arizona at approximately 4:25pm on Friday, 6 June, 2008. We arrived at the Base Camp at approximately 6pm and went about setting up camp. Once the tent was up, Kyle and Preston went exploring the area finding a pile of turkey bones. They had followed the trail from Campsite 1 approximately 100 yards where they located the bones and feathers of a turkey. The odd thing was the bones were picked clean much like when someone sits down and eats a Thanksgiving dinner. A bear, cat, or any other predator, will beat bones and all.

BUTT PRINT CENTER WITH TURKEY FEATHERS
IN UPPER LEFT CORNER

TURKEY KILL NUMBER 2

The reported back, and we decided to set up our first game camera on these remains hoping whatever would return and we would get a picture of the creature. While setting up the camera, we found two more turkey kills. Again, bones were all intact and not chewed.

Meanwhile I was working on getting my equipment working. Night vision scopes and motion sensors needed batteries, Game cameras needed SD cards, etc. A final check and they were ready to go.

I walked up to where the turkey remains were located and set the camera looking up the small game trail leading to the camp.

I set up the motion sensors so we could tell if anything was coming down the trail into our camp. Unfortunately, the only 9 volt battery I had available was too week to operate the sensor with any distance. It was virtually useless. We would have to do without the early warning it would give us if our turkey eating visitor decided to come into our camp.

I came back to camp and put two propane lanterns hanging from tree limbs to light up the campsite for the night. They were positioned as to leave no dark zones around the tent and cooking area.

As I finished placing the last propane lantern, Kyle and Preston returned to camp and started working on dinner. That is when the first rock came down from the sky just missing the hood of my truck. It landed with a very definite thud, and was about the size of an golf ball. It was not totally dark, but we were losing light so I fired up the lanterns. We looked around for the source of the flying rock, but were unsuccessful.

Just before dark we heard Alex Hearn squawking on the walkie-talkies. He was in his car headed for our camp and trying to reach us to see if we were already in camp. A short time later he arrived at the campground. When Alex arrived we showed him around the area and went to the first turkey kill to show him our camera set up. Afterwards, we returned to camp for dinner. It was good and tasty.

After dinner, Alex started setting up his truck for sleeping. He was going to spend the night in his tuck. As he was headed for the vehicle a second rock came crashing through the trees landing just short of him. It rolled to a stop. The rock came from uphill.

When the rock fell, we all froze. Instinctively we were listening for any clues of an attack. Movement on the hill, told us we were not alone. We were being watched. Of course we had a few tricks up our sleeve. We broke out the night vision scopes and began to study the hill side above us. Slowly, we began to venture away from the circle of light around our camp. We found nothing.

Being the first time we had ever encountered these circumstances we retreated back to camp when we didn't see anything. We were standing around the fire when the first vocalization was heard. It sounded like a whoop type call. It was not too far to the west of our camp. It sounded several times and each person in the camp got to hear it. This was not a pack of coyotes or wolves. It definitely was not an elk. We all agreed on this.

The vocalizations ceased, and we were talking about what we had heard. Alex picked up a large solid stick and took a whack at a tree. A few moments later, he did it again but this time three quick hits.

We listened for a few moments and suddenly the vocalizations retuned. This time there multiple calls. It sounded like two to the west of us coming our way. Then another group sounded from the South. We were getting quite excited about this when the third group chimed in from the North east. We were surrounded. The calls lasted for several minutes and ceased.

Things in camp began to return to normal. We had just had some very interesting events and our senses were pretty much on over load. Eventually, we one by one dropped out and went to bed.

About an hour before day break we (in the tent) awakened to a strange whistle. It was like nothing we had heard before. It sounded as if someone was whistling bird calls. We lay in our beds whispering to each other trying to figure out what would be making that kind of a sound. Eventually, the whistling moved around our tent and faded in the distance.

Day break came none too soon. It was good to be able to see the things around us. Breakfast came and went and we were off to check our game camera on the turkey kill. There were no pictures. The camera had not been tripped. But, the camera was working perfectly as evidenced by the picture of Kyle and Preston checking the counter on the face of the camera.

Our objective for the day was to go to where Susan had the tent collapsed down on her. She was sure it was not a bear. Her comments were, "Bears don't have hands, and they don't smell like that."

We headed up the road to the area above the water falls. On our way up we noticed a mine shaft on the side of the road. It was an old Uranium mine, and still had the radiation hazard sign posted at the entrance. Of course, none of us had any inclination to see if we would glow.

OLD URANIUM MINE WITH HAZARD SIGN

We finally reached the campsite. The first thing we noticed was how dense the growth was in the area. Ferns were up to our waist. The trees were very thick. We found a small trail going down the hill side to a creek. We began to notice the amounts of broken limbs with leaves laying on the ground. It was almost as if someone had been harvesting the smaller branches with leaves. We found several piles of these limbs under tree limb shelters. The shelters were smaller trees pulled over and wedged under a much larger tree limb creating a lean-to shelter. The limbs with leaves were all stacked under the lean-to with the broken (trunks side of branch) facing up hill creating a bed.

PRESTON SITTING IN THE FERNS

Looking around we began to find some partial foot prints. Nothing worth casting, but there was evidence of a lot of activity in the area. We worked our way up stream on the creek and eventually came out on the road. But, just before we got to the road we found a rain gage that had been totally bent to pieces. It looked as if someone had tried to take it apart with a very big hammer.

BROKEN RAIN GAGE

RAIN GAGE CLOSE UP

We topped out on the road and began to head back to our vehicles. We heard a truck coming, and sure enough it was the local Forest Ranger. He stopped to talk to us for a few minutes because he was curious about my truck. My truck had some magnetic signs on it displaying pictures of the Mogollon Monster and some big feet in the windows.

We explained to him why we were there and what we had been doing. He just smiled at us and said he had seen nothing strange, but had seen some scat that he had not been able to identify. He went on to say that there was a big brown bear in the area, but no one had seen it in about two years. In fact, no one had sighted any bears at all. However, there was a black panther ranging on a mountain to the south of us. So he cautioned us to be careful. He departed down the road.

As we were talking, Kyle was looking at something he had found o the side of the road not 20 yards away from where Susan had previously had her tent squashed. It was a huge footprint. We couldn't believe what we had found.

We counted the toes and discussed casting the print. None of us had any experience, but Alex had come prepared with footprint casting materials. We took our first footprint casting. (Photos of the 19 inch footprint are shown in Susan Farnsworth's story)

When we returned to our trucks, I was in for a surprise. A flat tire. I was very lucky to have a spare tire that was pumped up and ready for service. After fixing the tire, we prepared to leave back to civilization. Our time had run out.

We headed back home. On the way down the hill, Ranger Dave pulled us over. He seemed very excited, and he wanted to talk to us about our sightings—not knowing about our latest find. He told us more about the scat and where it had been found. The location was not too far from where we had recovered the footprint.

Ranger Dave went on to tell us about Crazy Larry, a hermit that lived in the local caves in the area. He wanted us to be careful not to mistake him for a bigfoot. We kind of laughed and showed him the footprint. We asked him if the hermit could leave a footprint like what we had. Dave got excited. He had never seen a print like this one. He was totally amazed. He told us that there was many stories of the area about lost Indians, a jaguar, but no bigfoot. He was now a believer. Eventually we had to say good bye, and we headed for home.

We cleaned the print and got to see our success. All we had to do was spray off the plaster with the garden hose and this is what we come up with. It is not an overstep bear print. You can clearly see the toes are in a slant. Bear toes form a V with the middle toe being the leading toe. The other bear toes slant away from the middle toe. Besides, there are no bears in Arizona with a 11 inch paw (compensating for the foot ending at the arch) being 9 inches wide. Arizona has no grizzlies, no Kodiak , or no polar bears.

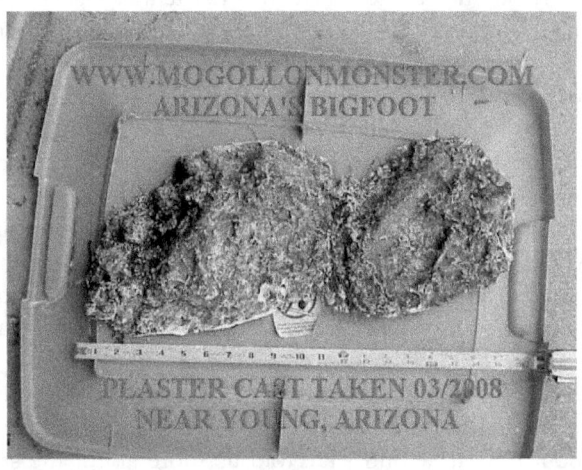

EXPEDITION 2 by Mitch Waite

Expedition 2: 19-21 June, 2008 Preston Smith, Mitch Waite, Alex Hearn and Tom (last name not given)

The purpose of this trip: Our objectives were to follow up on the Hot zone, and to place game cameras with bait. We also wanted to try to pull some fur using duct tape stung from trees and bushes in that area. We also wanted to interview more Forest Service Rangers about their knowledge of the area.

Preston and I departed from Mesa about 4:30pm. Our plan was to meet Alex Hearn from the Arizona Crypto-zoological Research Organization (AZCRO) and Tom at the campsite for our base of operations.

We arrived at the campsite just before dark only to find the entire camp ground had been taken by the Boy Scouts. There were tents everywhere. We found a small clearing near the campground entrance that was just big enough for our tent. Most of the evening we listened to the scouts run up and down the road shouting at each other. I guess they were all very excited about being in the outdoors. We listened to them play steel the flag, and flash light tag and playing in the creek.

Preston and I retired early an laid on our cots listening to the occasional shout and pranks being played over in the campground. About 11pm we began to hear the whistle from we had heard while we were on expedition 1. It came down off the mountain past our camp and stopped just prior to the first campsite in the campground. We could hear the scouts asking each other about the whistle. Then the cadence of the whistle got very fast. Same tune, but almost like it was enticing the scouts to come out and follow it.

I got up and could see the flashlights coming down the trail as they were chasing the whistle. It lead them up the small game trail past the turkey kills (locations we found during expedition 1). And finally both scouts voices and the whistle faded in the distance.

The next morning Preston and I were up and fixing breakfast when we saw the first scout vehicle pull out of the campground. It was if they a were in a hurry. On a Thursday, they were leaving. We thought that to be very odd. We had anticipated they would be staying over the weekend, but that was not the case.

Preston and I went fishing in the creek not too far from the campground. We did fairly well and caught 6 trout. We decided to keep two of the trout and use them for a bait/duct tape trap that we would set up later. Eventually we worked our way up the creek to the Lookout tower.

OUR CATCH OF THE DAY

We stopped and talked to Ranger Red. He was a very likable person and told us all about the history of the peak on which the tower stood. And he invited us up to take a look from the top. But, every time we asked him about bigfoot sightings, he would just shake his head and politely change the subject. Then just before we left, he commented that the ranger in the tower to the north of us had seen one. He had been riding a horse and it went bonkers. The ranger looked up just in time to see the bigfoot running up the mountain. He said it would reach up and grab the pine trees pulling himself up the mountain as it ran up the hill. He had never seen anything like it.

After the story we said good bye and headed back to our camp. We were expecting Alex and Tom to be there when we got back, but there was no one there. It was a strange and eerie thing when we drove into the campground. All the scouts were gone. Not a trace of them. We decided to take our pick of the newly available campsites. We had the whole campground to ourselves.

About an hour later, Alex and Tom arrived in the camp. We went through our introductions and set about making camp and dinner. Alex had just came from his car and was passing in front of my truck when a rock came crashing down from above. It caught our attention so we moved under the trees in case more rocks came our way. Nothing further happened until the next morning I noticed a big hole in the side of my trailer and a large stick approximately feet long laying next to the trailer.

TRAILER BEFORE ROCK THROWING INCIDENT

HOLE MADE BY STICK HITTING THE TRAILER

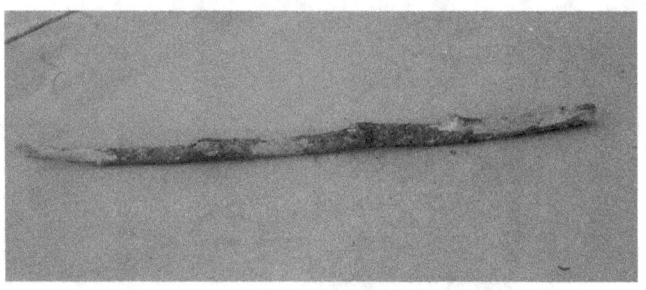

STICK FOUND LAYING NEXT TO TRAILER

The large end of the stick fits the hole in the trailer. It appears it was thrown much like a javelin. This factor has caused some disagreement as to if a Bigfoot could actually throw a stick like a spear.

There are some bigfoot researchers who claim Bigfoot has no opposable thumbs, and therefore can't throw a spear.

Consequently, they tell me the stick had to be thrown by a feral human such as a Hillbilly. However, we have never found Hillbilly tracks. Just tracks which resemble what most believe to be Bigfoot. It wasn't until the next year I was able to obtain photos of hands working on my game cameras, and a mud print to verify the photos were correct. It appears that even if they had no thumbs, they could still grasp a stick and throw it like a spear.

Both Alex and Tom were anxious about going to the site where Susan had her tent squished. This area became known to our group as the HOT ZONE. But, we took time out to do a walk around the camp. We showed them the two turkey kills from the previous expeditions. We found no foot prints other than humans (Boy Scouts). Breakfast was ready and waiting so we returned .

We made our way up the rugged road to the Hot Zone. Once there we went down to where we had found several trees that had been woven into shelters which we dubbed habitats. We found a tall tree and strung a sting over a limb and tied it to one of our trout. We hoisted it into the air to where it was about twelve feet in the air. Then we placed strips of duct tape all round to where if something reached up for the trout, it would lose some hair on the duct tape. We placed two game cameras on the area.

ALEX INSPECTING TAPE FOR HAIR

ONLY THE FISH HEAD REMAINED FROM THE
TROUT WE USED FOR BAIT

HAIR ON THE DUCT TAPE

SETTING UP A SECOND DUCT TAPE HAIR TRAP
AND CAMERAS

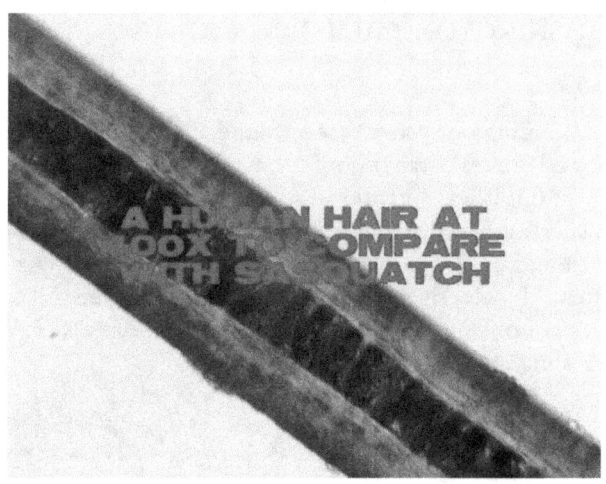

THIS IS ONE OF MITCH'S HAIRS FOR COMPARISON

HAIR FROM DUCT TAPE UNDER 400X
MAGNIFICATION. THIS HAIR MATCHES HIGH
PROBABLILITY BIGFOOT HAIRS FROM
WASHINGTON, CALIFORNIA, AND NEW YORK

Working our way west we found a small game trail and followed it. We ran into 5 scat piles. It looked almost human, but it was way too large and it was very light weight. We later learned that this was the type of scat Ranger Dave had found. It was black, and consisted of mostly plant matter. Finely chewed with exception of some Elkweed leaves. Look up the medicinal value of Elkweed. Very interesting.

FIRST PILE OF SCAT

MITCH BAGS THE PILES OF SCAT FOR FURTHER
ANALYSIS

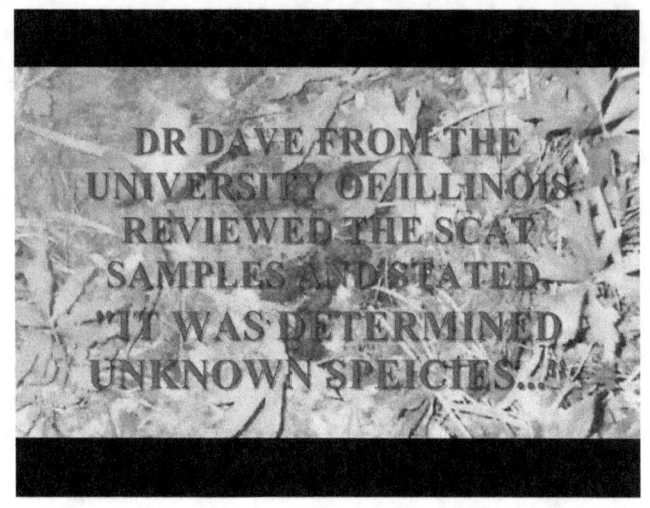

DR DAVE FROM THE
UNIVERSITY OF ILLINOIS
REVIEWED THE SCAT
SAMPLES AND STATED,
"IT WAS DETERMINED
UNKNOWN SPEICIES...

TOM HAD A BAG OF SCAT INSPECTED BY A
PROFESSOR FROM THE UNIVERSITY OF ILLISNOIS
DETERMINED THE SCAT TO BE FROM AN
UNKNOWN SPEICIES

ELKWEED

We worked our way back to the road and decided to go visit the Ranger up in the Lookout Tower. Red was there. He talked to us for quite some time until we mentioned the Mogollon Monster. He avoided the question and moved on to the history of the area. How the Army used the peak to signal the location of raiding Indians back in the 1800's. He also told us how 27 people lost their lives in a flashflood area by the water falls.

It came time to leave. Preston and I took the lead in my truck. Alex and Tom followed in their vehicle. On our way down the steep little two-rut road, I caught movement out of the corner of my eye. I turned to see the Mogollon Monster stand up in front of some rocks. I slammed on the breaks which caused a curious reaction from the Mogollon Monster. It was only 50 yards away and I could see it's face. Its countenance was "OH Crap! I've been seen!".

Preston thought I had seen something in front of us and was busy looking ahead. Alex and Tom had fell back too far in their vehicle to see what was happening other than I had slammed on my breaks.

CAVE WHERE BIGFOOT WAS STANDING

INSIDE CAVE HAS EXIT TO THE LEFT

INSIDE CAVE LOOKING RIGHT TO ANOTHER
EXIT

I took my eyes off the monster long enough to find the gear shift to slam it into reverse and look back. Within that short of time, it had covered nearly 30 yards and was entering a thicket of trees. It was gone!

As soon as Preston and Alex knew where I was looking they lit out on foot to give chase. However, it was too late. The Mogollon Monster had dropped over the ridge and disappeared. They never got a look at it. They soon returned to look at the cave. It turned out to be a shallow cave system with three entrances. It was a perfect shelter with more than one escape.

The interesting thing about the rocks is I had Preston stand where I had seen the Mogollon Monster. While he stood in the entrance of the cave, his head was about equal to the entrance rocks. These same rocks only came up to the Mogollon Monster's waist. Preston is 5'6". This means the monster was nearly 8 feet tall.

We were all kind of in shock. We just hung around discussing the event. Eventually, we had to leave, break camp and head back home.

EXPEDITION 3, JULY 3-4, 2008.
MITCH WAITE, PRESTON SMITH AND PHOEBE ADAMS.

Preston and I arrived early and set up our tent, outhouse tent and camp equipment. We decided to go fishing while we waited for Phoebe to arrive. We did fairly well. Enough for dinner. We arrived back in camp just before Phoebe.

She pitched her tent about 20 yards away from out tent and set up her Ramada over the picnic table. After dinner we kicked back and began to relax. The campsite at the other end of the campground was now taken. There were two adults and three kids. We could hear them playing in the creek and being kids.

Suddenly, there was an explosion. The other camp was setting off fireworks. This was totally illegal to do in the forest. Consequently, we figured all chances of hearing a bigfoot call or wood knocking was gone. They soon ran out of fireworks and things quieted down again.

During this time, Phoebe decided to try decorating a small pine tree with glow sticks and rings. She was hoping to attract the curiosity of any watching bigfoots. I placed one of my game cameras on it, and then retired to the campfire.

Phoebe broke out a boom box and was talking about playing some recorded from the internet bigfoot calls. We decided it couldn't hurt, but we were not expecting any results since the fireworks show. We tried several different calls, and whoops, and even beat on a tree with a large stick. But there were no returns. Phoebe tried one more which to me sounded like a drunken sumo wrestler cussing out his wife. We got a response. It wasn't what we hoped for but it was a response. We could hear one of the kids in the next camp scream, "MOM!"

We heard the adult reply, "Shut up and go to sleep!" We realized the adults were in the SUV and the kids were in a tent. Phoebe reached down and played the call again. This time all three kids responded, and we could hear the parents telling them to go back to sleep. By this time we were laughing. It was payback time for the fireworks. We laughed about it for a while and was amazed how the other bigfoot whoops and calls had no effect, but the weirdest call struck a tone of terror. Phoebe hit the boom box again. This time all we heard was four car doors slamming shut. The next morning they were gone. I kind of feel guilty for laughing so hard at the time, but we had the camp ground all to our selves the rest of the weekend.

Phoebe was first up the next morning, and she headed for the outhouse. I was next and went over to the picnic table to start breakfast. I went back to the tent to wake Preston when I saw it. On the ground not two feet from my tent door was a bare foot print. It was not huge, but was only 9 inches long and a good four inches wide. It was the weirdest foot print I had seen. We took pictures, videoed it and cast it in plaster. A second foot print was found in front of the shower tent.

CASTED FOOTPRINT IN FRONT OF TENT

The foot is nearly four inches wide in the ball of the foot, and shows a definite metatarsal break behind the ball of the foot allowing the rest of the foot to very lightly touch the soil making the heal to appear very small and off-set.

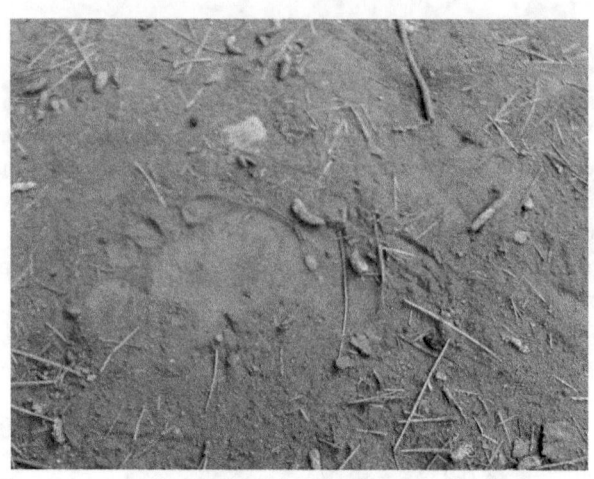

SMALL PRINT IN THE DIRT

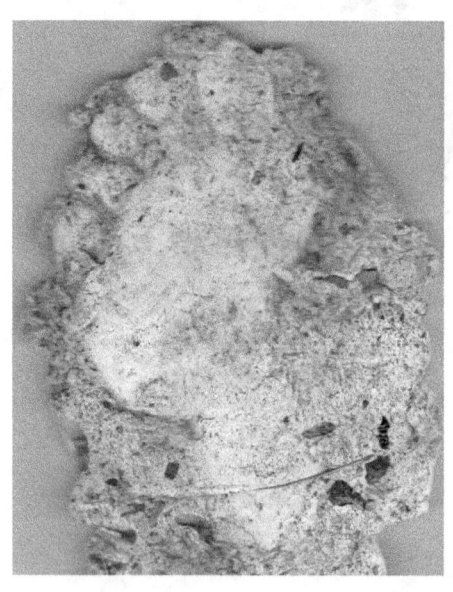

THIS FOOTPRINT SHOWS PECULIAR
CHARACTERISTICS.

Phoebe found a large ball of Spanish moss in front of her tent door. There was none growing in the camp ground. It had to be carried there. We kind of thought it was some kind of gift from our barefoot friend. Take time to look up the medicinal value of Spanish moss. A very interesting gift.

We believe we were visited early in the morning and they played "slap the bull" with our tents and left Phoebe a present. Some kind of thank you or peace offering.

After casting the footprint, we took Phoebe up to where I had seen the Mogollon Monster on the previous trip. We looked around a bit, and it looked as if someone was trying to make the cave look unused by piling all kinds of sticks in the entrances, that were not there previously. A bit strange.

Preston saw some rock bluffs above us and decided to go check them out. He found a series of small caves. One of which had a small wall built up in the entrance and a bed made out of leaves and grass inside. Phoebe and I followed and took several photos of the caves. We nicknamed this area the Mogollon Monster apartment complex. We now have game cameras on these caves to monitor any use.

It was time to head back. Our day was coming to an end, and we had a long drive to make it back.

CAVE 1 WITH GRASS AND LEAF BED INSIDE

PHOEBE CHECKING OUT SECOND CAVE

CAVE WITH WALL BUILT IN FRONT
NUT SHELLS FOUND INSIDE

EXPEDITION 4

EXPEDITION 4, 24-27 JULY, 2008. MITCH WAITE, PRESTON SMITH, AND ALEX HEARN

We arrived a Base Camp campground and set up camp. We were in Campsite 1. Nothing unusual happening other than a large rock hitting my trailer. It was a sizable rock, and my trailer was some distance from camp. It appears that my trailer was the intended target. The rest of the camp area is sheltered by trees, and throwing rocks at intended targets would not be very accurate due to the trees and leaves.

After setting up camp, we decided to go check our Duct Tape hair traps we had left three weeks earlier. We did use a trout as bait dangling from a string 12 feet up. The trout was surrounded with duct tape in hopes that if anything reached for the fish, any hair would get tangled in the sticky part of the tape. It had worked. Our trout was eaten completely with exception of the head. There were three hairs left on the duct tape. We bagged it all up for further lab analysis. (Photos displayed on pages 77 -82).

We separated and started searching the area for any further signs of activity. The first find was a pile of scat. I dawned my latex gloves and pulled out a zip-lock bag from my CSI backpack and proceeded to bag the scat four later analyses. It seemed to have the same plants and contents as the previous Scat we collected two months earlier.

A short time later we located a bed of sticks under a very large Spruce Tree. The Spruce tree was very thick and acted much like an umbrella. A 3 foot tall pile of sticks were neatly placed for bedding near the trunk of the tree. These sticks looks about the same age as the stick we found on the first expedition which still had their leaves on them. The earlier find of tree limbs was in the same area, but laying all over the forest floor as if they had been harvested out of the trees to be gathered later. All sticks were small with leaves. The tree limbs in the shelter were the same size which leads us to believe limbs from our first find was harvested to make the bed. However, not all the leaves had fallen off the branches.

In this same area we found many lean-to type shelters. Some seemed to be more sun shelters or hiding spots because they would not offer any protection from the wind.

SHELTER WITH PLACE TO SIT

We soon ran out of daylight and headed back to camp. During the evening, we could hear movement all around our camp. Preston and Alex headed up the mountain with some cameras and night vision, but were unsuccessful in catching any movement. It seemed as if they were being lured or played with. The noises would be just out of sight. At one time, I was able to catch some eye shine from our camp. They seemed a bit too red and too tall to be deer or elk.

The next day we decided to check out the area the locals called the tubs. It was a series of water pools and slides connecting them--A very nice place to spend a hot summer day swimming. Of course, it was crowded with hoards of humans.

We started back up the mountain to the falls. It was very impressive. 180 foot fall of water into a pristine pool at the bottom. We went on up to our hot zone and found a bed right next to the trail we had established when we were putting up the trout, duct tape trap. There was a huge bed of pushed down ferns and grass. Something has followed us back to the vehicles and then laid in the grass watching us getting ready to leave on the day before.

We decided to go up and visit the lookout tower on the highest peak. We stopped at the Forest Service Emergency cabin to look about for a while. Steve and I were walking down the cabin drive when he found a foot print much too large to be human. However, it was too washed out for any detail.

We continued on up towards the lookout tower, but stopped at the caves where I had spotted the Mogollon Monster on the first expedition. We went into the Apartment complex and took a good look at the leaf and grass mat matt built on the cave floor. We did not find any hair at that time.

We went on up to the lookout tower and found a Forest Ranger named Red. A very likable guy and had a lot of stories to tell. But, not any about the Mogollon Monster. That subject still seemed to be a bit off limits. However, I did notice that the Lookout tower was made with an iron door that separated the upper cabin area from the steps. It did look as if it was used quite regularly. I suppose that he closes the door at night when he gets ready to retire for the evening. I couldn't help but wonder if a Mogollon Monster could swing out around it and get on top of the deck. I wouldn't, but a primate might.

Our day was coming to end, and it was time to head back to civilization.

Some comments about the expedition: I searched for a lab to do some work on the hair and scat, but since the Bigfoot scam with the rubber suit in the freezer; none would do any work for me. Most said they would only for species identification if I was from a law enforcement or academic agency or organization. They would not do private work. Therefore, I bought my own laboratory quality microscope with a camera capability. As I removed the hair from the Duct tape, I thought for sure I would break it a dozen times because of the adhesive on the tape. Much to my surprise, the hair was extremely resilient and did not break. I took several pictures at 400x which clearly showed that the hair had no medulla and was not split in any way. I got on the net and Googled bigfoot hair samples. Low and behold, my picture matched the pictures from Maryland, Northern

California, and Oregon. All of these places claimed their hair to be authentic bigfoot hair. I compared the hair to bear, wolverine, mountain lion, raccoon, skunk, possum, deer, elk, and any other animal hair I could find on file with nothing close to matching my hairs. Proof enough that my hairs were from a bigfoot.

All expedition members were accounted for, and there were no other campers in the area. Access to the campground by automobile is from one direction only and is fully visible by the campsites in the area. There was no way someone could have came down the road and threw a large rock without being seen.

20APR09 EVENT
20-22 April, 2009

We arrived at the Base Camp campground at 10:30am and began to set up camp and make lunch. After camp set up we decided to take the cameras to the new location and set them out for overnight surveillance. On the way up, we stopped at several fishing holes on the creek to try our luck catching some trout. We were skunked with the exception of the fish that Sterling almost caught. It flipped off the hook before he could get it to the bank.

We arrived at the area where we would park the trucks and begin our journey into the dense forest. Shannon, Jerry, Sterling, and Ryan went West and Clay and I went East. After hanging my second camera, I caught movement of something across an opening to a section of thick brush. This was about the first time Clay smelled the foul odor. The chase was on. Clay was hot on the trail of the odor and sounds of movement through he brush. I was about seventy five yards behind him, but was managing to keep Clay in sight most of the times (my knees would not allow me to keep up).

I lost sight of Clay, but was following his tracks when I noticed a second set of footprints which paralleled Clay's prints, only these new footprints had toes. I figured Clay was tracking the new prints, but it just didn't seem right. I called out to Clay and he doubled back and found me studying the prints. He specifically stated these new prints were not there when he passed through that section of soft dirt. Then it hit him. He had become the hunted and not the hunter. We decided to return to the vehicles and meet up with Shannon's group. Then we could return to cast the best print in plaster. We made it to the vehicles with no problems.

CASTING OF FOOTPRINT FOLLOWING CLAY
HUMAN FOOT IN PHOTO IS A SIZE 9

CLAY PROUDLY SHOWS OFF FOOT PRINT
CASTING

CLOSE UP OF FOOTPRINT. IT LOOKS LIKE IT IS
MISSING A BIG TOE.

When Shannon returned to the vehicles we were waiting for him. He told us that he could hear wood knocking and it was no woodpecker. Sterling tried several different sticks on a pine tree to duplicate the sound. Eventually, he picked up a rock and hit the tree. Shannon's group stated that was the wood knocking sound they had heard. Shannon went on to say they all could hear someone talking in a low rough voice, and he assumed it was Clay and I talking and knocking on the trees. It was not us, and we told him so. Just to make sure it was not other humans, we drove up the road to the last possible parking place and looked for cars. There were none.

We doubled back in the vehicles to a place where we figured we could get to the footprint we wanted to cast. We had no problems relocating it, and soon had it covered with plaster. When the cast was hard, we washed it off in a nearby stream and found the foot only had four toes. The big toe was missing. To make sure we didn't mess up the big toe, we studied the left over print marks in the ground, and it confirmed there were only four toes. Could this be an injury, or was it born this way? The other toes were quite visible.

By this time, the shadows were beginning to lengthen and we decided to head back to camp for dinner. Along the way we stopped to gather fire wood at convenient places were trees had been pushed down over the road. Fortunately, the Forest Service had been in there and cut these trees into smaller sizes just right for the campfire. We had plenty of wood for two nights.

We arrived in camp and immediately set about fixing dinner. When we finished, we as a group decided to do some night vision work in a large meadow that had a spring and three apple trees. This would require us to go back to nearly where we were before. There was no moon, and therefore was a very dark night.

I stayed at the vehicles to listen for vocalizations, while the others hiked up the trail to the meadow to view it with night vision scopes. They were gone for about 20 minutes when I saw the first hint of flashlights piercing the trees and sky. They were returning to the vehicles in a hurry. When they reached us, Shannon told us they were doing good until he stepped on a large stick and frightened something in the bushes. It ran Away from them sounding like a "freight train". Shannon stated there were "a lot of animals" in the meadow but he could not tell what they were. When they jumped the freight train, it spooked them and they decided to return to the vehicles. Shannon remarked several times that it could have been an elk, but he couldn't tell.

SOMETHING JUMPED OUT OF THE BUSHES AND SUPRISES OUR RESEARCHERS. THEY NEVER GOT TO SEE WHAT IT WAS, BUT IT WAS BIG!

We walked down the road for a while looking for "eye shine", but did not see anything. We eventually decided to return to camp for the night.

We were sitting in camp with a nice fire talking about Mogollon Monster stuff when we heard the first scream. It seemed a long ways away, but it kept screaming as it got closer. All of us, admitted it was nothing we had heard before with the exception of myself. I had heard the same screams the year before in the very same campsite. We agreed it was not a coyote or wolf. More like howler monkeys. There seemed to be three of them. The screams continued to get closer and louder until they sounded as if they were on the top of the hill overlooking our campsite. By this time I had found my recorder and sound amplifier and was recording the cries. The cries continued for over 30 minutes and we decided to chase the sound down. We knew the main road cut back behind the screams so we decided to try to get a better location of the sound. We drove down the main road until we thought we were in the right area and stopped. We turned off the lights and killed the engine. We had passed it, and it was somewhere behind us. We turned around and found a small fire break road and turned in. We came to a clearing and we stopped, shut the engine off and turned off the lights.

THE AUTHOR RECORDING THE
VOCALIZATION EVENT

To our surprise, the next cries that came sounded like they were only a few yards away. I must admit, I was getting a bit spooked with the lights off in the pitch black thinking something might reach through the open window and drag me out of the car. Upon the next scream we began to exit the cars and proceed into the forest to track down the screams. Unfortunately, the car alarm went off on the other car when the doors opened. This silenced the screams. We waited and listened, but no more sounds. Eventually, we gave up and returned to camp. We had on sooner started warming our hands by the fire when the screams returned. They lasted for about five more minutes. Almost like they were defying us to come and chase them again.

Finally, it died down and the rest of the evening was quiet. It was over. Was it a challenge to us? Was it upset at us for disturbing it in the meadow a couple of hours earlier? Was it lost and trying to locate another of its kind? Consensus, was it was telling us it was his territory and for us to stay out. Of course, we didn't. The next morning we started out by going back to where we had parked the cars during the chase. We found no concise proof of anything. However, I did find many places where turkeys had moved the pine needles to create a dust bath. Could the screams have been to alert others where dinner was located?

After breakfast we headed back to the camera areas to check the photos. I got 10 pictures. We changed out the memory cards and reset the cameras. While we were doing this Clay found a large grey hair stuck to the camo on the camera. We later determined it to be a match (not in color but in size and composition) to the baby Bigfoot hair we got last year. Notice there is no medulla which means it is not human. Here are the pictures for them:

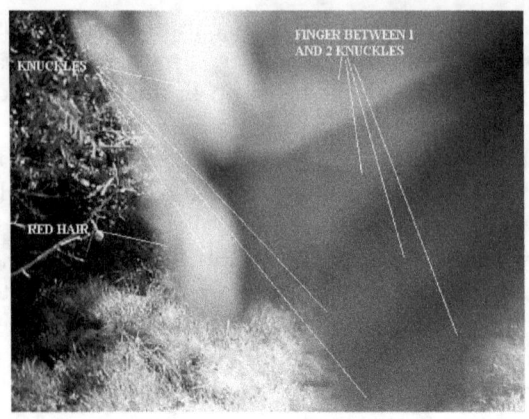

HAND REACHING FOR THE CAMERA

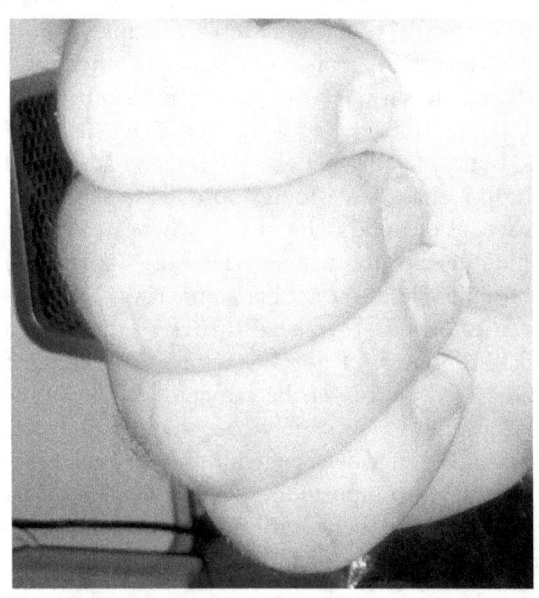

HUMAN HAND FOR COMPARISON

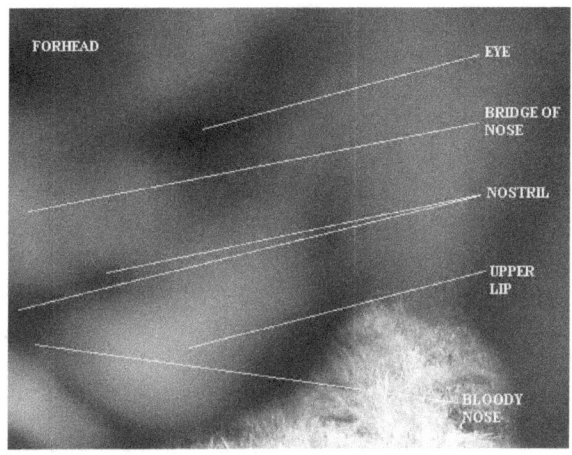

FACE OF THE OWNER

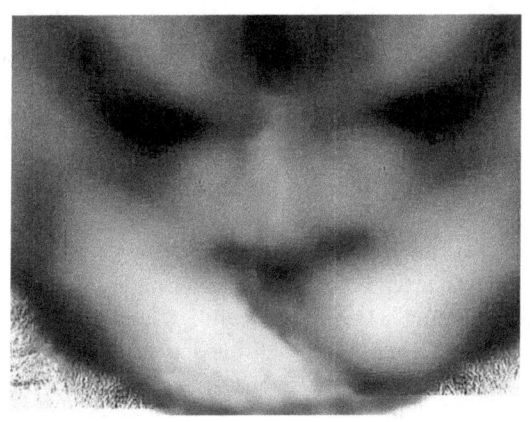

WITH A LITTLE PHOTO WORK AND SYMETERY
WE GET A LOOK AT A FULL FACE OF A BABY
MOGOLLON MONSTER

We then headed up the road to where I had spotted my first Mogollon Monster the year before. We spotted nothing, but the cave. We then went to visit a local ranch. It was a beautiful place with over 20 apple trees in some very nicely trimmed fields. Afterwards, we gathered more fire wood, because the wood we had before was consumed after the screams. They had kept the fire good and high all night long.

I decided to place a trail camera down at a permanent spring and cement trough. While heading that way we discovered a turkey kill. It had not been there the day before. Sometime while we were out of the camp someone or something had came down within 50 yards of our camp and plucked a turkey. There were no head, feet or guts. Therefore, we concluded it was not human. Everything was missing but the feathers. Then not more than 10 feet away from the feather pile we discovered some small bare foot prints. If it was a Bigfoot, it was a baby. It left three good prints in the wet sand. Could it be a baby playing in the creek while mom or dad plucked the turkey for dinner? Could the turkey have came from the area we found the dust baths? Were the screams maybe the celebration of a successful kill?

We cast the prints in plaster, but due to the wetness we would have to leave them in place over night.

The evening was very peaceful. No screams or out of place noises. I got some much needed sleep. However, our two young men spent the night watching over the camp and keeping the fire high.

The next morning, it was time to break camp. I had retrieve the cameras up near the meadow. Clay and I took this on early while the others slept. As we headed up the road, we came upon 5 turkey running up the road. They keep just in front of the truck for about 75 yards before they finally decided to get off the road and head up the mountain. It was a very rare sight indeed.

SIZE 9 BOOT COMPARED TO SMALL
FOOTPRINTS

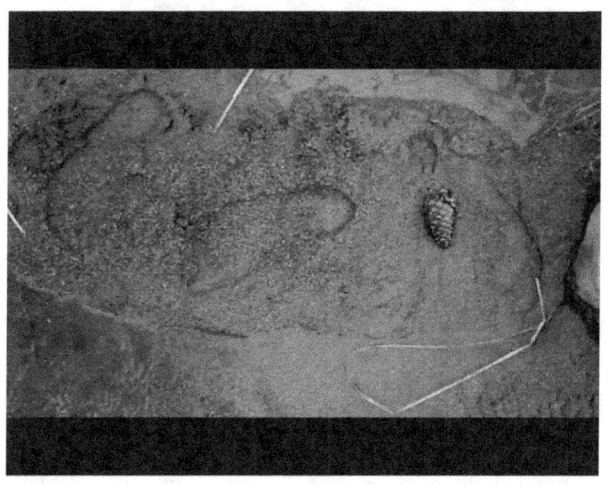

ALL THREE FOOTPRINTS

Clay and I retrieved the cameras with no out of the ordinary circumstances, and we returned to camp. On our way back there was another turkey standing in the road. It soon ran away, and we continued back to camp.

We started to break the camp down for us to return to civilization. But, I needed to retrieve my camera near the cement pond, and the three prints we had cast in the sand. We found the three prints. One of the prints had already been removed from the sand and was upside down in the water. Someone or something was interested in the plaster in the sand. We saw no other Bigfoot prints, I suppose it could have been a raccoon or skunk curious with the funny white rock in the sand. But then again, maybe it was something else that had moved the footprint. We retrieved the three prints and headed back to camp.

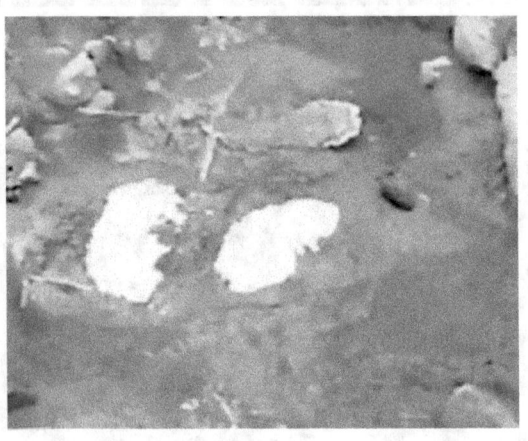

FOOTPRINT TURNED OVER

We finished breaking camp and started to leave. We spotted the orange on the table and thought we should retrieve it before the Forest Rangers found it and blamed us for baiting bears. We found the orange was squished and the slices removed from the peals leaving a round hull of peal. Very strange indeed.

The trip was over, it was a long, boring trip back to town.

15JUNE2009

Monday: We had some trouble with the trailer on the way up. We lost a wheel bearing which almost caused us to lose a wheel. I ended up having to put the trailer on some big rocks and pull the axle to take it back to town to have it fixed. My trailer is still up there in the mountains. I hope to get back up there to put it back together and bring in home in the next couple of days.

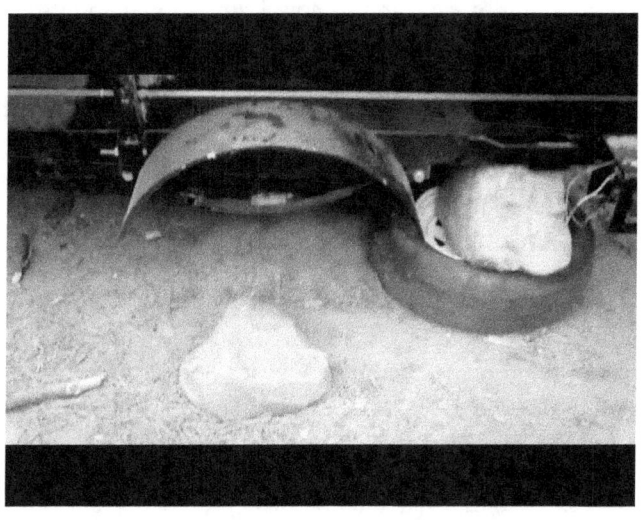

TRAILER BEARINGS OUT

After taking care of the trailer axle, Preston and I decided to fish the creek while David hiked some trails around the camp ground. David didn't have much to tell when we got back from fishing. However, I caught my limit of trout in about an hour. Preston caught a couple too.

While we were fishing, I was surprised by two dogs (what I believe to have been two Mexican Grey Wolves). They approached me in a very friendly way, and sniffed my hand all the while wagging their tail like a normal house dog would do. I was thinking they were just an off breed of Husky at first until I spotted the radio collar around the neck of the big one. Evidently, these two were raised by humans and were glad to see me. They showed no fear of me, and I did not think about them being wolves until I saw the collar. I had thought they were from a nearby camp. Preston and I got in the truck and traveled down the road to the next fishing hole and the two dogs followed. Eventually, they gave it up and disappeared. Now I am worried about them getting shot because of their friendliness. To describe them, they were bigger than a coyote but smaller than a German Shepard. They looked like a wolf with a slight yellow tint to their grey fur.

We fixed dinner and settled down next to the camp fire to relax. About 8 pm, David heard what he described as a roar. He said he didn't know what a bear would sound like, but it was a definite roar quite some distance away. About two minutes later all of us heard a scream. It was not human, but I couldn't say it was a Bigfoot either. No other out of the ordinary sounds were heard during the rest of the evening.

Tuesday morning I awoke to the sound of a whip-or-will call. Last year that sound usually preceded some kind of activity around the camp. But nothing happened. So I snoozed until the sun started peeking though the pine trees into our camp.

Our objective for the day was to retrieve the game cameras in the field. We were headed up the dirt road to the camera area, when we saw something move down the side of the road towards the creek. It was a very large cat. We pulled up to where it went off the road, and David got a look at its face. I saw it as it crossed the creek below to the other side. It was the biggest bobcat I had ever seen. This might have explained the roar and scream from the night before. It disappeared into the foliage.

Nothing happened while we gathered the cameras. No sign, no foot prints, no scat, no wood knocking, vocalizations, nothing.

We returned to camp, ate dinner and enjoyed a nice quiet evening around the fire. No screams. nothing.

We got up early Wednesday morning to head back to civilization. We took everything out of the trailer we could and put it into the truck. Sadly I left my trailer behind in the campground parking lot. I hope it is still there when I get the axle back to it.

Note: The trailer was fine, and I replaced the bearings and installed the axle. It is working fine.

Some of the photos retrieved from the cameras were very intriguing. One photo had a hairy something across the bottom of the photo, and the rest of it was washed out in a light purple glow with the exception of what appeared to be an eye. Victor Oropezo ran the photo through some color filters and was able to construct a face. Here is the series:

ORIGINAL HAIRY ARM PHOTO

FIRST PASS THROUGH FILTERS, A FACE
BEGINS TO APPEAR

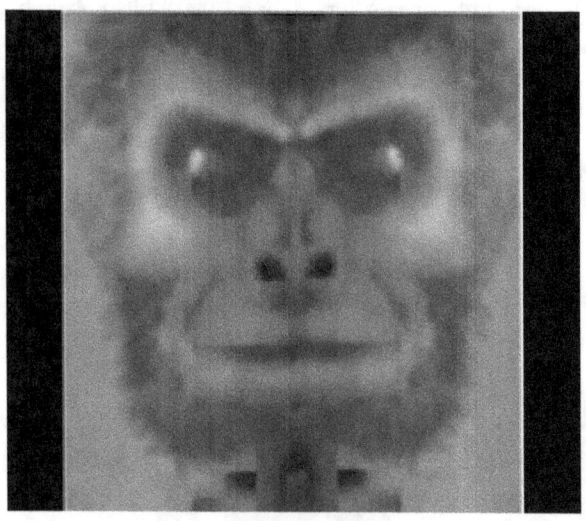

A COUPLE OF MORE PASSES AND A FEW
TOUCH-UPS AND THE PHOTO FILLS IN TO SHOW
THE FACE OF AN ADULT MOGOLLON MONSTER

We now have a composite face to study. Yes, it is photo-shopped and computer generated, but none the less, it gives us an idea as to what our Mogollon Monster looks like as an adult. Our work and research has just begun. We really know very little about this creature.

WE HAVE BEEN LIVING WITH THEM ALL ALONG, CONVINCED THEY WERE SOMETHING ELSE...NOT WILLING TO BELIEVE—Mitch Waite